# Energy-Efficient Electric Motors and Their Applications

## Second Edition

# Energy-Efficient Electric Motors and Their Applications

Second Edition

Howard E. Jordan

Plenum Press • New York and London

Library of Congress Cataloging in Publication Data

Jordan, Howard E.
Energy-efficient electric motors and their applications / Howard E. Jordan.—2nd ed.
p. cm.
Includes bibliographical references and index.
ISBN 0-306-44698-7
1. Electric motors—Energy consumption. 2. Electric power—Conservation. I. Title.
TK2781.J67 1994
621.46—dc20 94-5315
CIP

The first edition of this volume was published by Van Nostrand Reinhold in 1983. It is now out of print.

ISBN 0-306-44698-7


A Division of Plenum Publishing Corporation
233 Spring Street, New York, N.Y. 10013



Printed in the United States of America

# Preface

The worldwide energy shortage has had a significant impact on the design and application of electric motors. The opportunity is now available, as never before, to select motors which can result in substantial energy savings. Wise selection and use of energy-efficient motors not only supports our national energy conservation goal but also provides attractive financial rewards to the motor user. This book is dedicated to providing the technological background and understanding for selecting and using electric motors to accomplish their mission with minimum energy consumption.

Many motor manufacturers now offer energy-efficient lines of motors with the result that motor buyers now have two distinct categories of products from which to select: standard and energy-efficient motors. However, not all applications can profit from the use of energy-efficient motors. In fact, there are some applications in which an energy-efficient motor will perform less satisfactorily than a less efficient type of motor. In certain applications, the energy-efficient variety will not even provide the necessary driving capability. The plant engineer, engineering consultant, installing contractor, or other faced with the problem of selecting a motor or replacing an existing motor is left in a quandary. Should he use an energy-efficient motor or not? If he does, will he save money? What other characteristics will be affected? People faced with these decisions are usually not specialists in electric motor technology. This book provides a guide to answering these questions in a form usable to engineers and contractors, as well as to others who do not specialize in motor technology.

A U.S. Department of Energy study has identified the 5–125 horsepower range of induction motors as the segment in which almost one-half of all motor-drive energy is used. This segment, therefore, offers the greatest potential for energy savings. This book is primarily directed toward polyphase induction motors, but some discussion is included on other types, namely single-phase, excited-field synchronous, and synchronous-reluctance machines. An entire section (6.4) has been devoted to the permanent-magnet synchronous motor in this Second Edition. This type of motor has become increasingly popular in recent years. Section 6.4 explains why and provides information on the types of permanent magnets used, the performance characteristics of these machines, and the applications where permanent-magnet motors are in use.

Chapter 7 is devoted to adjustable-speed motor drives. Adjustable-speed drives are widely used to control speed, tension, position, etc. for industrial processes. In addition to serving a broad range of control applications, adjustable-speed drives can also provide an energy-savings benefit relative to constant-speed motors for certain applications. These energy-savings considerations are discussed in Chapter 7.

Chapter 7 also describes several of the more important types of electronic controls used to provide the adjustable-speed function. This discussion starts with the phase-controlled dc drive and then describes several types of inverters and brushless dc drives. The chapter provides a technical background which is useful in deciding whether or not to use adjustable speed or constant speed. If adjustable speed is the choice, the technical information on various types of drives will be useful in selecting the proper type.

The era of the energy-efficient motor has arrived. Energy consumption is a parameter which needs to be considered in each motor selection and application situation. This book has been written to help in making this decision wisely.

Howard E. Jordan

*Cleveland, Ohio*

# Contents

# 1

# Introduction

## 1.1. Need for Energy-Efficient Motors

During the period when energy was plentiful, electric motor technology was directed toward providing durable products with extended service life as economically as practical. The Arab oil embargo of the 1970s and the ensuing awakening to a worldwide energy shortage ushered in a new era for the electric machinery industry. The impact of these events on electric motors and what can be done to conserve our dwindling energy sources through judicious selection and application of electric motors are the subjects of this book.

It is interesting to observe the progress of motor technology over the past several decades. Tracing the history of the NEMA 445 frame size, one finds that in 1930 the rating housed in that frame was 50 hp at 1800 rpm. Today the 445T frame general-purpose ac motor is rated at 200 hp. This dramatic advance was made possible through a combination of improved design, materials, and manufacturing processes. In addition, today's 200-hp motor provides longer service life in much more severe environments than any of its predecessor designs.

These actions have been anti-inflationary and have contributed in a major way to our nation's industrialization. The relatively inexpensive energy conversion from electrical to mechanical form provided by electric motors has made possible many of the advances in industrial technology that we enjoy today.

Now, with energy efficiency a prime goal, a new dimension has been added to motor technology. Product lines of energy-efficient motors have emerged and motor buyers have distinct categories to select from: standard and energy-efficient motors.

## 1.2. Energy Usage Profile

A look around our homes and any industrial or commercial plant would easily convince one that a substantial quantity of energy must be consumed by motor-driven devices. The amount of energy, relative to that consumed by automobiles, heating, lighting, etc., is not generally well-known. The Department of Energy concluded a study on electric motors and pumps in February, 1980, and issued a report which provides some insight into the portion of energy consumed by electric motor drives. The data on energy usage cited in this chapter are based on the Department of Energy report.[1]

In 1977, electric motor drives accounted for 18.3% of the total national energy usage. For comparison, automobiles used 13.5% of the nation's energy. Approximately 58% of the electric energy generated in the United States is utilized by electric motor-driven equipment. The other 42% is used for heating, lighting, etc. Manufacturing accounts for 45.2% of this 58%; other big users are transportation, communication, and utilities, which, as a group, consume 20%, and commercial applications, which account for 18%. Although the typical household contains several electric motor-driven appliances such as a refrigerator, washer, dryer, and electric razor, the household accounts for only 4.8% of the total electrical energy generated for motor drives.

Based on this revelation, it is clear that a significant portion of the nation's energy is consumed by motor-driven equipment and that a potential for important energy conservation exists by focusing on motors and the equipment they drive.

## 1.3. Types of Electric Motors and Their Energy-Saving Potential

There are two broad categories of electric motors: alternating current (ac) and direct current (dc) motors. Since ac power is generated and distributed by electric utilities, it is to be expected that the large majority of electric motors are of the ac variety. Direct current motors are primarily found in traction applications and in industrial applications where controlled speed is required by the process being driven. The dc motor is ideally suited for situations in which speed control is required because of the ease with which its speed can be controlled by varying its armature voltage, field excitation, or a combination of the two.[2] Complex drive systems consisting of many dc motors, their solid-state power supplies, and their electronic regulators provide precise regulation and coordination between different driven sections of machinery. Many of today's industrial processes are highly dependent on these dc drives.

Traction drives also use dc motors partially because of their ease of control but also because the on-board power source is frequently dc, usually a battery. Although dc motors serve an undeniably important function in the nation's industrial and transportation activities, their total drive energy utilization represents slightly over 1% of the energy utilized by the entire motor population in the United States.[1]

There are many varieties of ac machines, but they can be grouped into two general categories, synchronous and asynchronous. The synchronous speed of an ac motor is determined by the number of poles for which it is wound and the frequency of the power source in accordance with the equation

$$\text{Synchronous rpm} = \frac{120 \times \text{frequency}}{\text{No. of poles}} \tag{1.1}$$

Synchronous motors run at an average speed in accordance with Eq. (1.1), while asynchronous motors operate at a speed different from synchronous speed.

The horsepower of synchronous motors is usually very large or very small with relatively little usage in-between. For example, electric clock motors are synchronous motors, but their shaft horsepower is minute. Industrial processes requiring precise synchronization between sections of the process sometimes use synchronous machines but the bulk of this usage falls into the 1 hp and lower category.

The region between fractional horsepower and the very large horsepowers is dominated by the induction motor, which is an asynchronous machine. The horsepower size above which the synchronous motor becomes competitive with an induction motor depends on the speed and the need of the plant for a unity power factor or even a leading power factor to correct for other lagging power factor loads on the plant. A complete discussion of power factor and its effect on energy consumption is included in Chapter 5. A synchronous machine's power factor is controlled by its excitation. By overexciting the machine it can be operated at a leading power factor and this can be used advantageously to offset lagging power factor loads on the same electrical system. Totally, synchronous machines account for less than 1% of the motor-drive energy used in the United States, according to the 1980 Department of Energy report.

The ac polyphase induction motor dominates the energy usage spectrum insofar as electric motors are concerned. The energy usage of this category, specifically ratings of 5.1 hp and greater, is 93% of the total electric motor usage. Induction motors rated 5 hp and lower, including the fractional horsepower category, account for approximately 5% of the total motor energy utilization in the United States. This may seem surprising because most of the motors we encounter in everyday life are in the fractional and subfractional horsepower sizes. In fact, the fractional and subfractional horsepower motor population exceeds the above 5-hp sizes of motors by over 10 to 1 but, nevertheless, the preponderance of energy usage is in the 5-hp and above category. An explanation for this incongruity is the intermittent usage pattern of most household appliance motors. The Department of Energy report further segments the polyphase induction motor category and identifies the 5.1 to 125 hp range as utilizing nearly 50% of the total motor energy and thus it is this category that has the greatest potential nationwide for energy savings.

Mindful of this energy usage profile, this book concentrates on the polyphase induction motor. However, the other major types of motors are also discussed, and in Chapter 7 the application of variable-speed drives for energy conservation purposes is considered.

## References

1. "Classification and Evaluation of Electric Motors and Pumps." Washington, D.C.: U.S. Department of Energy, November, 1980.
2. Schieman, R.G., Wilkes, E.A., and Jordan, H.E. Solid state control of electric drives, *Proceedings IEEE,* Vol. 62, No. 12, December, 1974.

# 2

# Technology of Energy-Efficient Motors

## 2.1. Energy Usage in Motor Applications

Electric motors function as energy converters changing electrical to mechanical energy and are one of the most efficient means of energy conversion known. Although some losses do occur in the conversion process, the majority of the electrical energy which enters the motor is converted to mechanical form and delivered to the load. Consider Fig. 2.1 which shows a 50-hp motor driving a pump of equivalent rating. The power distribution is shown in the figure. The electrical input into the motor is 41.2 kW and, of this amount, 37.3 kW is converted and delivered to the pump. This 37.3 kW is expended by the pump in circulating the fluid that it is pumping plus, of course, any losses occurring in the pump. The motor losses account for the difference between the incoming electrical power and the converted power delivered to the pump which, in this case, is 3.9 kW. The motor's efficiency is 90.5%.

Two important observations can be made from this example of a typical motor application. First, the power "consumed" by the motor in this example is only 3.9 kW or 9.5% of the total power delivered to the motor by the electrical power system. This power, which supplies the motor losses, is dissipated in the form of heat. The remaining 37.3 kW is converted and delivered to the motor's load, the pump. A popular misconception has arisen from energy usage data quoted in some of the technical press. One of the common conclusions often quoted

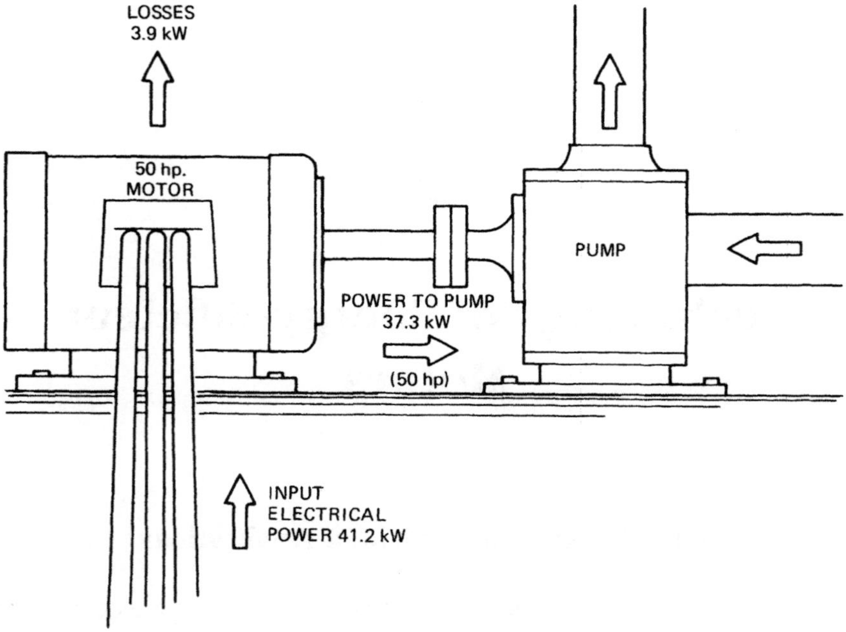


**Figure 2.1.** Power distribution in motor driven pump system.

is that electric motors "consume" 58% of the electrical energy generated in the United States. This is completely erroneous. The motor consumes only a small portion of this 58%, that portion which supplies the motor losses; the rest of the energy is converted to supply the load.

Another important observation is that good energy management requires that the entire motor–load system be evaluated to minimize energy waste. Clearly, the best one could do by concentrating on the motor alone would be to eliminate the entire 3.9 kW of motor loss but this is only 9.5% of the total system power. This would require a 100% efficient motor which is impossible, but application of an energy-efficient motor in this instance would provide substantial savings. However, it is also clear that there is a potential for energy saving by looking beyond the motor to the driven equipment. Improving the efficiency of the driven equipment can often provide energy-saving rewards equal to or greater than improving the motor efficiency. Both the motor and the driven equipment should be critically examined for potential energy-savings opportunities. NEMA Standards MG10[1] and MG11[2] provide some very good guidance for applying polyphase and single-phase motors, respectively, with energy efficiency as a goal.

## 2.2. Five Components of Motor Losses

A cross-sectional view of an induction motor is shown in Fig. 2.2. The induction motor is the type which is used in the largest percentage of motor-driven applications from an energy usage standpoint.

The view shows the principal components of the motor which are:

1. Stator laminated steel core
2. Stator windings
3. Rotor core comprised of steel laminations and cast aluminum bars and end rings
4. Rotor fan blades which are an integral part of cast aluminum bar and end ring structure
5. Shaft
6. Bearings
7. Frame
8. Brackets
9. External fan
10. Fan cover

The rotor structure for induction motors of medium size usually consists of a stack of steel laminations which are held together by cast aluminum bars and end rings. It is this bar and end ring assembly which carries the current that is induced in the rotor. The bars extend through slots in the rotor steel laminations and are short-circuited at either end by the circular end rings. The cast aluminum bar and end ring assembly resembles a cage used for exercising pet squirrels; hence, the name squirrel-cage rotor has been adopted for this type of construction. The axial blades extending off the end rings are fan blades used to circulate cooling air inside the motor.

There are five components of loss in an induction motor, namely:

- Primary $I^2r$
- Iron

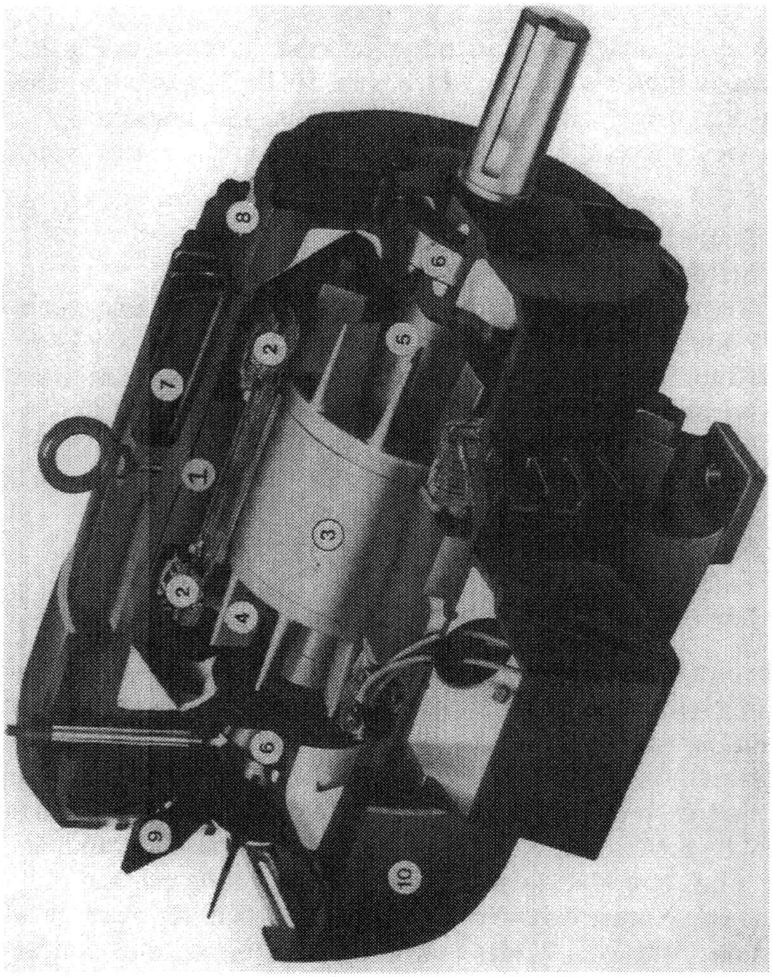


**Figure 2.2.** Cross-sectional view of a Reliance Electric Energy Efficient XE Motor. 1, stator laminated steel core; 2, stator windings; 3, rotor core; 4, rotor fan blades; 5, shaft; 6, bearings; 7, frame; 8, brackets; 9, external fan; 10, fan cover. (Courtesy of Reliance Electric Company.)

- Secondary $I^2r$
- Friction and windage
- Stray load

The primary $I^2r$ is the ohmic loss caused by current passing through the stator winding. Most motors use copper conductors, except for small-horsepower machines where aluminum wire is sometimes used. When copper windings are used, there is little more that can be done, from a practical standpoint, to increase the winding conductivity. However, increasing the cross-sectional area of copper to reduce the primary resistance is an effective means of reducing the primary $I^2r$ loss and this method is used in nearly all energy-efficient motor designs. Adding copper cross-sectional area, of course, adds pounds of copper and cost to the motor.

The iron loss occurs in the stator laminated steel core because of the combined effect of hysteresis and eddy current losses. These losses are a result of the magnetic field in the stator core which oscillates at line frequency. Iron losses vary as some power, greater than 1.0, of the magnetic flux density in the iron. Therefore, iron losses can be reduced by reducing the flux density and this can be achieved by adding length to the stator core. Thus, most energy-efficient motors will contain more iron and more copper than a standard motor of the same rating.

Another effective means of reducing iron losses is to use a better grade of steel for the laminations and also to reduce the thickness of the laminations. The quality of a magnetic steel relative to its loss production for a given flux density is affected by the silicon content and the processing of the steel during manufacturing. Increasing silicon content, up to some practical limit, reduces the watts per pound loss, as does annealing the steel. Reducing the thickness of the laminations effects a reduction in the eddy current component of loss. All of these techniques are used to reduce the iron losses in energy-efficient designs.

The same scenario might seem to be equally applicable to the rotor, but this does not turn out to be the case. The secondary $I^2r$ losses similar to the primary $I^2r$ losses are the ohmic losses in the rotor conductors. However, the secondary $I^2r$ losses are not controlled so easily by simply adding more conductor material to the cage structure. The motor's starting torque is proportional to its secondary resistance. Since the motor must meet some minimum specified value of starting torque, usually the NEMA specified value for Design B motors,[3] there is a limited range through which the secondary resistance can be varied. Therefore, the starting performance specifications tend to limit the amount by which the secondary $I^2r$ loss can be reduced to favor efficiency. Frequently, the secondary $I^2r$ loss will be the least affected loss in changing from a standard design to an energy-efficient design.

Design A motors can usually be designed to be more efficient than a Design B motor of equivalent rating. A Design A motor does not have a specified NEMA limit for locked-rotor current, and this allows more freedom to vary the secondary resistance. The net effect is that some increase in motor efficiency can be achieved at the expense of higher inrush current. In applications where the power system is not adversely affected by higher inrush current demands, the compromise to permit Design A could result in some energy savings.

Fundamental frequency iron losses are not present in the rotor laminations because the rotor's magnetic field oscillates at slip frequency, usually only a few cycles. Iron losses at low frequency are normally considered to be negligible. There are some high-frequency flux pulsations in the rotor, particularly near the surface at the air gap. These fall into the general category of losses classified as stray load loss, the fifth component of motor losses discussed in more detail below.

The friction and windage losses are those associated with rotation and subtract from the net power delivered to the load. In a fan-cooled motor such as the one shown in Fig. 2.2, the major component of friction and windage loss is usually the power taken by the external fan. There is no interchange of air between the inside and outside of a fan-cooled motor. All of the heat, therefore, must be removed by convection and the temperature drop associated with this process is reduced by passing high-velocity air across the motor's frame.

Additional windage loss is created by the rotor fan blades which circulate air within the motor, and, of course, there is some friction loss in the bearings. Energy-efficient motors have less losses to be dissipated than a standard motor and, therefore, smaller external fans can frequently be used as less airflow is required. The smaller fans have a favorable effect on friction and windage loss and, therefore, contribute to increased motor efficiency.

Stray-load loss is defined as the difference between total motor losses and the sum of the other four losses, namely, primary $I^2r$, iron, secondary $I^2r$, and friction and windage. Stray-load losses are the most elusive of the five losses within a machine as they are caused by several different factors. A complete discussion of all of the causes and cures for stray-load losses could be the subject of another book in itself. Only a brief discussion of some of the major elements making up the stray-load loss will be considered here.

The air-gap flux density is usually considered to be sinusoidal, but, in fact, it contains many imperfections related to the slotting of the stator and rotor as well as saturation effects. These imperfections result in high-frequency currents in the rotor bars and high-frequency iron loss in the rotor and stator teeth near the air gap. These harmonic rotor currents provide little useful torque, but cause an ohmic loss in the rotor bars. Iron parts such as baffles located close to the magnetic field of coil end turns can be responsible for additional iron losses which are classified as stray-load loss.

The control of stray-load losses must be accomplished by a combination of design and careful manufacturing practices. These losses are an important component of the overall motor loss, and reduction of this component is essential to the truly energy-efficient motor.

Table 2.1 shows a typical loss distribution for a 50-hp standard motor and an energy-efficient motor of the same rating. The last column of the table tabulates the loss improvement for each of the five losses. It is evident that the major loss improvements were in the stray-load loss, primary $I^2r$, and iron loss, with friction and windage making some contribution; and in this case, the secondary $I^2r$ loss actually went up slightly. The net reduction in losses for the energy-efficient

**Table 2.1. Typical Losses—50-hp, Four-Pole Motor**

| | Standard motor 50 hp losses (kW) | Energy-efficient motor 50 hp losses (kW) | kW loss improvement |
|---|---|---|---|
| Primary $I^2r$ | 1.319 | 0.911 | 0.408 |
| Iron | 0.725 | 0.180 | 0.545 |
| Secondary $I^2r$ | 0.646 | 0.668 | (0.022) |
| Friction and windage | 0.373 | 0.281 | 0.092 |
| Stray load | 0.852 | 0.299 | 0.553 |
| Total | 3.915 | 2.339 | 1.576 |

motor in this example was 40% compared with the standard 50-hp machine. This is a substantial loss reduction when one considers that the energy-efficient motor still provides industry standard locked-rotor torque, breakdown torque, and locked-rotor current, so that it will be completely interchangeable with the standard motors.

## 2.3. Calculation of Loss Reduction

The motivation for applying an energy-efficient motor is to accomplish the delivery of the required mechanical energy to the load with a minimum of wasted energy in the form of motor losses. Thus, it is motor losses that are the central area of concern in evaluating alternate motor choices for a particular application. A review of the definitions of power, energy, and efficiency is presented first in this section as an introduction to calculation of losses, loss reduction, and economic payback.

Power is defined in physics as the "rate at which work is done."[4] The units for electrical power are watts (W) or kilowatts (kW; 1000 watts). In the United States, motor output ratings are usually expressed in horsepower (hp), a unit nearly equal to 746 W. In many foreign countries, horsepower is not used and motor output ratings are expressed in kilowatts.

Energy is the product of power and time, and if power is constant in time,

$$W = Pt \tag{2.1}$$

where $W$ = energy, $P$ = power, and $t$ = time.

Electrical energy is usually measured in kilowatt-hours (kWh). If a constant power usage of 1 kW is maintained for a period of 1 hour, then the energy consumption is 1 kWh, and if the electric utility charges 5¢ per kWh the cost of the energy used would be 5¢.

The efficiency of any device is given by

$$\text{Efficiency} = \frac{\text{Output}}{\text{Input}} \tag{2.2}$$

An alternate form of (2.2) is given by

$$\text{Efficiency} = \frac{\text{Output}}{\text{Output} + \text{Losses}} \tag{2.3}$$

which focuses on the losses as being the subject for consideration in choosing between different motors, or any other type of equipment for that matter, when energy conservation is the issue.

In performing energy usage calculations, usually the known quantities are the horsepower output rating of the motor and the efficiency at rated load. The power input in kilowatts is given by

$$P_i = \frac{\text{hp} \times 0.746}{\eta} \quad \text{in kW} \tag{2.4}$$

where $P_i$ = input power, hp = horsepower output, $\eta$ = efficiency expressed as a decimal value, i.e., percentage value divided by 100, and the motor losses (in kW) are calculated as

$$L = \text{hp} \times 0.746 \left[\frac{1.0}{\eta} - 1.0\right] \tag{2.5}$$

Of course, if operation at other than the rated load point is being considered, the horsepower at the operating point and the efficiency associated with it must be used in (2.4) and (2.5).

Consider a 10-hp motor with an efficiency of 90%. Using (2.5) the motor losses are

$$L = 10 \times 0.746 \left[\frac{1.0}{0.90} - 1.0\right]$$

$$= 0.829 \text{ kW}$$

To realize a loss reduction of 20%, the machine's efficiency must be improved to 91.8%. However, if the same 10-hp rating had an initial efficiency of only 72%, its efficiency would have to be improved to 76.3% to realize the same 20% reduction in losses. To continue this example, suppose the initial efficiency for the 10-hp motor was 97%, then a 20% loss reduction would require improving the efficiency to 97.6%. The losses for each of these examples as calculated from (2.5)

**Table 2.2. Differences in Motor Losses for Changes in Efficiency**

| | | 10-hp output | | |
|---|---|---|---|---|
| Efficiency % | Input kW | Output kW | Losses kW | Loss difference % |
| 72.0 | 10.361 | 7.460 | 2.901 | 20.1 |
| 76.3 | 9.777 | 7.460 | 2.317 | |
| 90.0 | 8.289 | 7.460 | 0.829 | 19.7 |
| 91.8 | 8.126 | 7.460 | 0.666 | |
| 97.0 | 7.691 | 7.460 | 0.231 | 20.8 |
| 97.6 | 7.643 | 7.460 | 0.183 | |

are tabulated in Table 2.2. While the efficiency range, 72% to 97.6%, is an unrealistically large range for a normal 10-hp motor design, the example illustrates the point that it is the difference in losses that one needs to focus on in choosing between alternatives in motor selections, not the difference in percentage points change in efficiency.

Generally, efficiency values for motors increase with their horsepower rating. A typical efficiency for a 500-hp motor is 95% and for a 5-hp, 85%. Let us assume that energy-efficient versions of these two motor ratings were available and that their respective efficiencies were 95.5% and 87%. The kilowatts saved by choosing the energy-efficient designs are shown in Table 2.3. Clearly, it would take many 5-hp motors to produce the equivalent kilowatt savings realized by only one 500-hp motor even though its percentage point efficiency improvement was only 0.5%.

Two additional equations which are useful in evaluating changes in motor losses are (2.6) and (2.7):

$$L_s = \text{hp} \times 0.746 \left[\frac{1}{\eta_1} - \frac{1}{\eta_2}\right] \tag{2.6}$$

**Table 2.3. Comparison of Kilowatts Saved**

| Horsepower | Output kW | Efficiency % | Input kW | Losses kW | Savings kW |
|---|---|---|---|---|---|
| 500 | 373.0 | 95.0 | 392.6 | 19.6 | 2.0 |
| 500 | 373.0 | 95.5 | 390.6 | 17.6 | |
| 5 | 3.730 | 85.0 | 4.388 | 0.658 | 0.101 |
| 5 | 3.730 | 87.0 | 4.287 | 0.557 | |

where $L_s$ = savings in motor losses (kilowatts), hp = horsepower output, $\eta_1$ = efficiency expressed as a decimal for the lower efficiency value of the two being considered, and $\eta_2$ = efficiency, higher value. The difference in motor losses expressed as a percentage of the motor losses associated with the less efficient motor is

$$\% \text{ change in losses} = \left[\left(\frac{1}{\eta_1} - \frac{1}{\eta_2}\right) \Big/ \left(\frac{1}{\eta_1} - 1.0\right)\right] \times 100 \tag{2.7}$$

The relationship between typical motor efficiencies and watts loss versus horsepower is illustrated in Fig. 2.3.

## 2.4. Economic Payback Calculations

The savings in motor losses can readily be equated to a dollar savings in energy cost. Two additional factors must now be considered, namely the hours of operation and the cost of electrical energy. The dollar savings per year (or month) are then given by

$$S = L_s \times H \times C \tag{2.8}$$

where $S$ = savings per year (or month) in dollars, $L_s$ = savings in motor losses from Eq. (2.6), $H$ = operating time per year (or month) in hours, and $C$ = cost of electrical energy in dollars per kWh. Equations (2.6) and (2.8) can be combined into a single equation for calculating savings:

$$S = \text{hp} \times 0.746 \left[\frac{1}{\eta_1} - \frac{1}{\eta_2}\right] H \times C \tag{2.9}$$

Equations (2.8) and (2.9) apply to motors operating at constant load. If the load varies, the savings must be calculated piecewise for each operating interval over which the load can be considered to be constant and the savings for each operating interval summed to give the total annual savings.

A simple payback computation yielding the time to pay back the

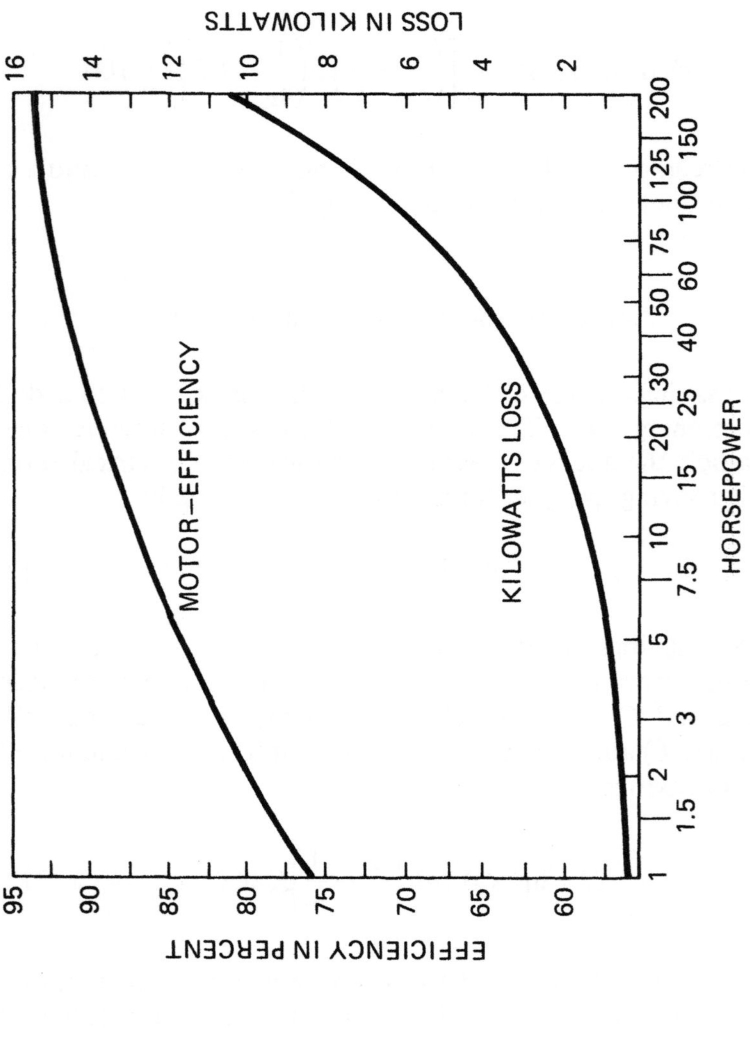


**Figure 2.3.** Efficiency and kilowatts loss versus horsepower.

cost differential between two motors with differing efficiencies is made using

$$\text{PBP} = \text{CD}/S \tag{2.10}$$

where PBP = payback period in years (or months), $S$ = savings from Eq. (2.9), and CD = cost difference between the two motors in dollars.

Using the example from the previous section of the two 10-hp motors, one with 90.0% and one with 91.8% efficiency, dollar savings per year and the payback period are calculated below.

Dollar Savings and Payback Calculation Example

| | Motor No. 1 | Motor No. 2 |
|---|---|---|
| Horsepower | 10 | 10 |
| Efficiency | 90.0% | 91.8% |
| Hours of operation per year | 6000 | 6000 |
| Motor cost | \$400 | \$500 |
| Energy cost per kWh | \$0.05 | \$0.05 |

$$S = 10 \times 0.746\left[\frac{1}{0.900} - \frac{1}{0.918}\right]6000 \times 0.05$$

$$= \$48.76$$

$$\text{PBP} = \frac{500 - 400}{48.76}$$

$$= 2.05 \text{ years or } 24.6 \text{ months}$$

The calculation of a simple payback using (2.10) ignores a very important fact in our economic life, the time value of money. We are all familiar with the fact that if we borrow \$6000 from a bank for 3 years to buy a car, we will have paid the bank back considerably more than the original \$6000 by the end of the 3 years when the loan is completely paid. The same principle applies to spending additional money to buy an energy-efficient motor over a standard motor. The impact of the "cost of money" can be taken into account by using the present-worth factor (PWF) in calculating the payback period.[5]

The PWF is given in (2.11) and is used to find the present worth, $P$, of some future amount, $F$:

$$\text{PWF} = \frac{1}{(1+i)^n} \tag{2.11}$$

where $i$ = interest rate expressed as a decimal and $n$ = number of interest periods

$$P = (\text{PWF}) \times F \tag{2.12}$$

The calculation of the payback period from the preceding example is repeated in Table 2.4 to include the effect of interest on money. In Table 2.4, the interest rate is assumed to be 12% per year or 1% per month and the calculation is made with the appropriate present-worth factor applied each month to the energy savings that accrue during the month. The effect of considering the interest on money in calculating the payback period is to lengthen it from 24.6 months in the first example where no interest was considered to nearly 29 months.

Another investment analysis technique used to evaluate potential energy-efficient motor purchases is life cycle savings, LS. In this method the cost per kilowatt of loss over the anticipated life of the motor, KWC, is calculated by (2.13) and this value is used in (2.14) to calculate the life cycle savings:

$$\text{KWC} = H \times C \times n \tag{2.13}$$

where KWC = cost of a kilowatt of loss, $n$ = years of motor life, and $H$ and $C$ are as defined for (2.8)

$$\text{LS} = \text{hp} \times 0.746 \left[\frac{1}{\eta_1} - \frac{1}{\eta_2}\right] (\text{KWC}) \tag{2.14}$$

If the life cycle savings, LS, exceeds the initial cost differential between the two motors, then purchase of the more efficient motor is indicated.

Using the data from the previous example and assuming the life of the motor to be 10 years, the life cycle savings can be calculated using (2.13) and (2.14):

**Table 2.4. Discounted Investment Payback Calculation**

| | Motor No. 1 | Motor No. 2 |
|---|---|---|
| Horsepower | 10 | 10 |
| Efficiency | 90.0% | 91.8% |
| Hours of operation per year | 6000 | 6000 |
| Motor cost | $400 | $500 |
| Energy cost per kWh | $0.05 | $0.05 |

$$S\text{(per month)} = 10 \times 0.746\left[\frac{1}{0.900} - \frac{1}{0.918}\right]\frac{6000}{12} \times 0.05$$

$$= \$4.06$$

Interest rate = 12% per year

= 1% per month

$$\text{PWF} = \frac{1}{(1 + 0.01)^n}$$

| Month | Energy cost savings | PWF | Discounted energy cost savings | Accumulated discounted energy cost savings |
|---|---|---|---|---|
| 1 | $4.06 | 0.9901 | $4.02 | $4.02 |
| 2 | " | 0.9803 | 3.98 | 8.00 |
| 3 | " | 0.9706 | 3.94 | 11.94 |
| 4 | " | 0.9610 | 3.90 | 15.84 |
| 5 | " | 0.9515 | 3.86 | 19.70 |
| 6 | " | 0.9420 | 3.82 | 23.52 |
| 7 | " | 0.9327 | 3.79 | 27.31 |
| 8 | " | 0.9235 | 3.75 | 31.06 |
| 9 | " | 0.9143 | 3.71 | 34.77 |
| 10 | " | 0.9053 | 3.68 | 38.45 |
| 11 | " | 0.8963 | 3.64 | 42.09 |
| 12 | " | 0.8874 | 3.60 | 45.69 |
| 13 | " | 0.8787 | 3.57 | 49.26 |
| 14 | " | 0.8700 | 3.53 | 52.79 |
| 15 | " | 0.8613 | 3.50 | 56.29 |
| 16 | " | 0.8528 | 3.46 | 59.75 |
| 17 | " | 0.8444 | 3.43 | 63.18 |
| 18 | " | 0.8360 | 3.39 | 66.57 |
| 19 | " | 0.8277 | 3.36 | 69.93 |
| 20 | " | 0.8195 | 3.33 | 73.26 |
| 21 | " | 0.8114 | 3.29 | 76.55 |
| 22 | " | 0.8034 | 3.26 | 79.81 |
| 23 | " | 0.7954 | 3.23 | 83.04 |
| 24 | " | 0.7876 | 3.20 | 86.24 |
| 25 | " | 0.7798 | 3.17 | 89.41 |
| 26 | " | 0.7720 | 3.13 | 92.54 |
| 27 | " | 0.7644 | 3.10 | 95.64 |
| 28 | " | 0.7568 | 3.07 | 98.71 |
| 29 | " | 0.7493 | 3.04 | 101.75 |

$$\text{KWC} = 6000 \times 0.05 \times 10$$
$$= \$3000/\text{kW}$$
$$\text{LS} = 10 \times 0.746 \left[\frac{1}{0.900} - \frac{1}{0.918}\right](3000)$$
$$= \$487.58$$

Since the life cycle savings of $487.58 exceeds the $100.00 price premium, the energy-efficient motor is a good investment.

Life cycle savings calculations can also be made to include the effect of the cost of money. The cost of a kilowatt hour of energy is modified by the "present-worth factor for an equal payment series"[5] which accounts for the fact that the savings accrue at some future time. This factor is given by

$$\begin{array}{l}\text{Present-worth factor for}\\ \text{an equal payment series}\end{array} = \frac{(1+i)^n - 1}{i(1+i)^n} \tag{2.15}$$

Equations (2.13) and (2.14) have been restated below to include the effect of interest and the example for the 10-hp motors has been recalculated using a 12% annual interest rate.

$$\text{KWC}' = H \times C \left[\frac{(1+i)^n - 1}{i(1+i)^n}\right] \tag{2.16}$$

$$\text{LS}' = \text{hp} \times 0.746 \left[\frac{1}{\eta_1} - \frac{1}{\eta_2}\right](\text{KWC}') \tag{2.17}$$

$$\text{KWC}' = 6000 \times 0.05 \times \left[\frac{(1+0.12)^{10} - 1}{0.12(1+0.12)^{10}}\right]$$
$$= \$1695.07/\text{kW}$$

$$\text{LS}' = 10 \times 0.746 \times \left[\frac{1}{0.900} - \frac{1}{0.918}\right](1695.07)$$
$$= \$275.50$$

Although the life cycle savings have been reduced considerably by including the present-worth factor, the savings still exceed the initial cost differential between the two motors.

Payback calculations are sometimes carried out with much more complexity than the ones shown here. Such items as investment tax credit, income tax, salvage value, etc., can be included in the financial analysis for an investment decision. The methods shown here give several alternative means for determining if the purchase of energy-efficient motors are a desirable investment for a particular application. If the results of this type of analysis are favorable and the size of the purchase warrants it, then a more rigorous financial analysis may be made to serve as a basis for the final investment decision.

## References

1. NEMA Standards Publication No. MG10, "Energy Management Guide for Selection and Use of Polyphase Motors." Washington, D.C.: National Electrical Manufacturers Association, 1983.
2. NEMA Standards Publication No. MG11, "Energy Management Guide for Selection and Use of Single-Phase Motors." Washington, D.C.: National Electrical Manufacturers Association, 1977.
3. NEMA Standards Publication No. MG1, "Motors and Generators." Washington, D.C.: National Electrical Manufacturers Association, 1993.
4. Resnick, R., and Halliday, D. *Physics for Students of Science and Engineering.* New York: John Wiley & Sons, Inc., 1960.
5. Thuesen, H.G., Fabrycky, W.J., and Thuesen, G.J. *Engineering Economy.* Englewood Cliffs, N.J.: Prentice-Hall, Inc., 1977.

# 3

# Selection and Application of Energy-Efficient Motors

## 3.1. Applications Which Can Benefit Most from Energy-Efficient Motors

There are certain characteristics which can be used to identify applications where energy-efficient motors should be used. These characteristics are identified in this section.

Long, continuous operating periods at or near rated load is a feature that suggests the use of an energy-efficient motor. Since operating time per year is a factor in calculating the dollar savings [see Eq. (2.8)], the savings will become larger as the operating time increases. A good example is a ventilating fan which runs 24 hours per day, 7 days per week. An improvement in motor efficiency is almost certain to result in a significant energy savings in this type of application. On the other hand, a motor driving an electric door opener may only operate a few times per day and then only for a few seconds during each operating period. While an energy-efficient motor may indeed reduce the power consumption during the time the motor is in operation, the total energy savings over a year will still not be very remarkable.

Attention should be focused on the applications where the highest losses occur. Figure 3.1 displays the relationship of typical motor efficiencies and kilowatts loss versus output horsepower. From this figure, the efficiency of a 100-hp motor is 91.8% and its losses are 6.664 kW when operating at rated load. Now, if this motor were replaced with an energy-efficient motor having an efficiency of 93.3%, approximately 20% less losses, the power savings would be 1.307 kW.

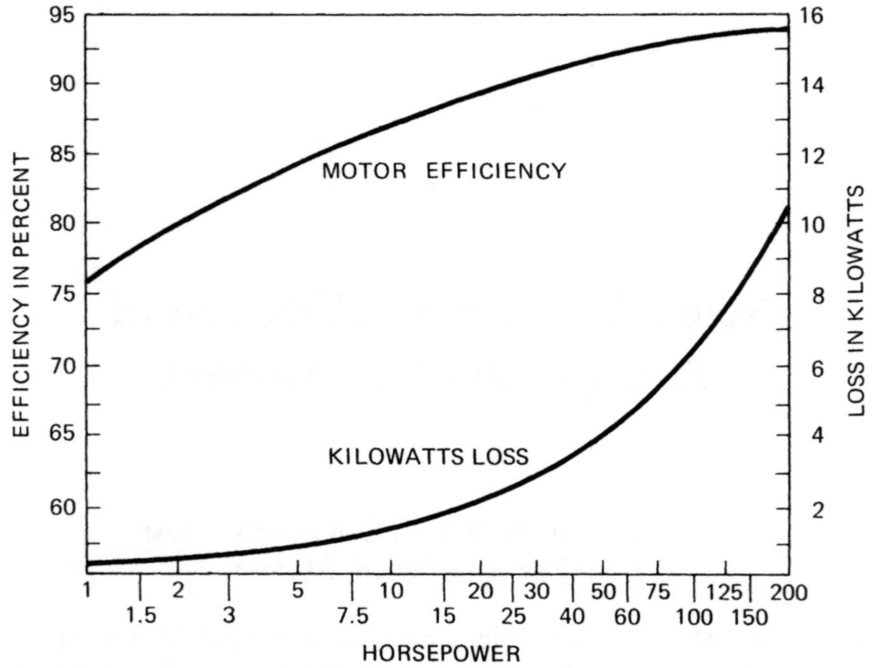


**Figure 3.1.** Motor efficiency and kilowatts loss.

On the other hand, a 3-hp motor has an efficiency of 82.0% and, according to the figure, has losses of0.491 kW. Replacing the 3-hp rating with an energy-efficient design at 85%, again approximately a 20% loss reduction, results in a savings of 0.096 kW. Although the efficiency change in the case of the 3-hp motor is 3 percentage points compared to 1.5 points for the 100-hp rating, clearly the potential for energy savings is much greater with the 100-hp motor. For the best return on investment, the kilowatts loss should be calculated in the various parts of the plant or building being considered and then attention focused on the areas where the losses and operating hours are the largest.

## 3.2. Matching the Motor to the Load

It is important to match the motor to the load properly if the full benefits of an energy-efficient motor are to be realized. Figure

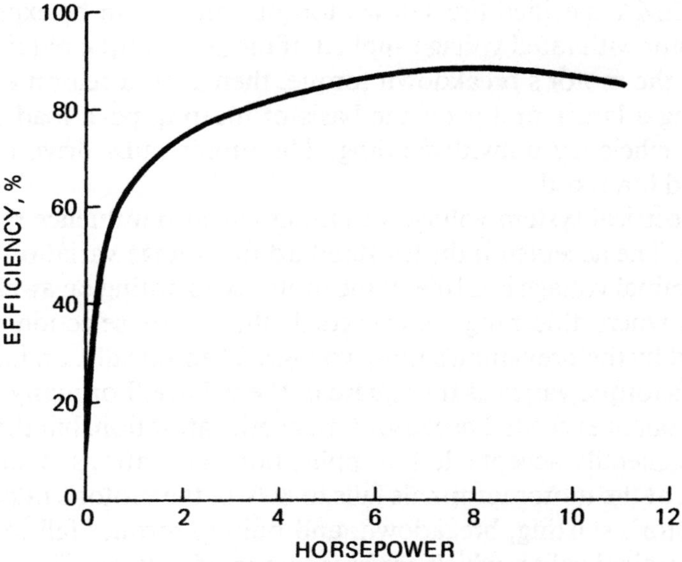


**Figure 3.2.** Percent efficiency versus load for 10 hp motor.

3.2 displays the efficiency versus horsepower load characteristic for a typical 10-hp motor. Curves for other horsepower ratings are similar. The motor efficiency remains reasonably constant over the range from 1/2 to full load and many motors actually have slightly better efficiency at the 3/4 load point than they do at full load. These statements are generalities, of course, and there will be some designs which depart from the rule, but most NEMA Design B motors have similar characteristics. The general shape of the efficiency versus load characteristics is applicable to both standard and energy-efficient motors.

As a consequence of this efficiency versus load characteristic, it is desirable to select a motor which will operate most of the time in the region between 3/4 and full load. When the load is unknown, it is often the practice to oversize the motor to ensure satisfactory operation. In some cases there may be no other choice, but it is evident from Fig. 3.2 that operating the motor lightly loaded negates some of the benefits that can be realized by applying a properly sized motor. If possible, determine the load torque and select a motor which will match the torque at close to the motor's nameplate rating.

Some caution must be observed in following these guidelines. If the load is subject to wide fluctuations, the motor must have sufficient breakdown torque to carry it through the peak load torque condition.

The NEMA specified breakdown torque values[1] can be expected of the motor with rated voltage applied. If the peak torque requirements exceed the motor's breakdown torque, then a valid reason exists for selecting a larger motor on the basis of meeting peak load requirements, efficiency notwithstanding. The motor must drive the load, first and foremost!

Electrical system voltage variations can also influence motor selection. The accepted industry standard for voltage variation around the nominal voltage is $\pm 10\%$. If the motor is operating on an electrical system where this range is exceeded, the motor selection may be affected by the prevailing supply voltage. Theoretically, an induction motor's torque varies as the square of the voltage. For many motors, the exponent exceeds 2 because of magnetic saturation, but the square law is generally acceptable for application calculations. Thus, if the voltage at the motor terminals falls to 80% of the motor's rated value, the motor's starting, breakdown, and pull-up torques fall to 64% of their nominal values which are based on rated voltage. The best solution in this case is to correct the voltage conditions so that the motor can be properly sized for its load and, therefore, operate at optimum efficiency. Departure from the rated voltage $\pm 10\%$ standard will usually result in a degradation in efficiency.

In an existing installation where the motor is being replaced by an energy-efficient design, the motor speed, horsepower rating, design type, etc., are established. However, in a new application where choices are available, there are some general guidelines which, if followed, will help to improve energy utilization.

For the same horsepower rating, two- and four-pole motors with synchronous speeds of 3600 and 1800 rpm, respectively, have higher efficiency than slower-speed motors. Of course, if a gear reduction is required, the overall efficiency must be considered, since the gear reducer will also have an efficiency which is related to the required speed reduction. The efficiency of motors increases with horsepower size, as is evident from Fig. 3.1, so if the choice is between a larger motor versus several smaller machines, energy usage favors the larger machine.

The synchronous speed and slip of an induction motor are defined by

$$\text{Synchronous speed} = \frac{120 \times \text{frequency}}{\text{No. of poles}} \tag{3.1}$$

$$\text{Slip} = \frac{\text{synchronous speed} - \text{operating speed}}{\text{synchronous speed}} \tag{3.2}$$

The losses in the rotor of an induction motor are proportional to the slip, and, therefore, low-slip motors are more efficient than high-slip motors. High slip may be required by the application but, if a choice is available, a normal slip design will have better efficiency.

NEMA has defined several design classifications for induction machines:

- Design A—Locked-rotor (starting inrush) current exceeds the value specified for Design B motors. The torque values will equal or exceed those specified for Design B. An increase in locked-rotor current will usually be accompanied by a reduction in slip, and, therefore, a Design A motor can be designed with better efficiency than a comparable Design B machine. A design trade-off of improved efficiency for increased inrush current is sometimes possible providing the electrical supply system is capable of withstanding the increased current demand during starting.
- Design B—Design B motors meet specified maximum values of locked-rotor current and minimum values of breakdown, locked-rotor, and pull-up torques.[1] General-purpose ac motors and most machines in the energy-efficient category are Design B machines. The speed–torque characteristics of Design A, B, C, and D machines are displayed in Fig. 3.3. Slip for Design B motors at rated load is less than 5%.
- Design C—This type of design is characterized by a higher locked-rotor torque than Design B. The locked-rotor current specification is the same as for Design B and this type of machine is used for applications where high breakaway torques are encountered in getting the load started. A typical application is a conveyor.
- Design D—Design D motors are high-slip machines and have a speed–torque characteristic which is distinctly different from the other design categories, as is evident from Fig. 3.3. They also exhibit higher locked-rotor torque values than Design B machines and these values are specified in the NEMA "Motors and Generators" publication.[1] They are used in applications where lifting, starting, or accelerating the load is a more de-

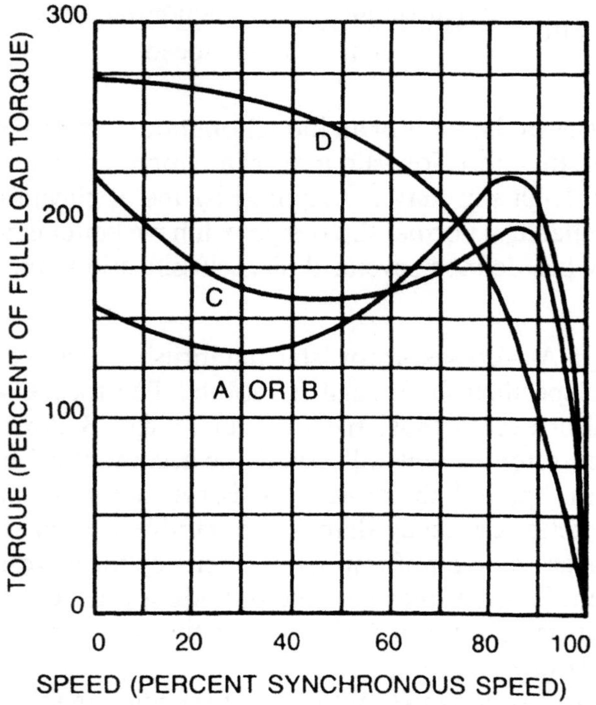


**Figure 3.3.** General shape of speed–torque curves for motor with NEMA designs A, B, C, and D. [This is Fig. 2-2 of NEMA Standards Publication MG 10-1983(R1988). It is reproduced by permission of the National Electrical Manufacturers Association which holds the copyright.]

manding feature of the application than running under full load. Slips for Design D motors at rated load equal or exceed 5%. The high slip associated with a Design D motor reduces its running efficiency, and, therefore, a Design D motor would usually not be selected for loads to be driven continuously near the rated horsepower output of the machine. Typical applications for Design D motors are punchpresses, cranes, and hoists.

## 3.3. Applications Which Do Not Benefit from Energy-Efficient Motors

The hours of operation are an important factor in deciding whether or not an energy-efficient motor should be selected. Section

2.4 gave methods for making a quantitative evaluation of the economic gains to be realized by applying a higher-efficiency motor and the hours of operation are included in all of the various methods discussed for making this evaluation. Highly intermittent duty applications such as valve operators, door operators, etc., are not prime candidates for energy-efficient motors.

Design D motor applications are also ones in which the energy-efficient motor will not find a place. Sometimes the load requires a higher starting torque than is available from a Design B or C motor and, therefore, Design D is the only choice. A less obvious case where Design D is the proper choice is one where a high-inertia load is accelerated frequently. The concept is best illustrated by the following example.

EXAMPLE: HIGH-INERTIA LOAD ACCELERATION CALCULATIONS. A 10-hp motor is required to accelerate a 50 $\text{lb}\cdot\text{ft}^2$ inertia in accordance with the following duty cycle:

a. Accelerate load
b. Run at rated load for 30 seconds
c. Coast to rest
d. Repeat a, b, and c 10 times per hour

Compare a Design B and Design D motor with respect to the energy consumed in performing this duty cycle.

Motor losses during a speed changing period are predominately $I^2r$ losses in the stator and the rotor. The secondary losses in an induction motor, $W_R$, can be calculated from Eq. (3.3) under the assumption that there is no appreciable retarding load torque during the acceleration. The more general case where there is a retarding load torque as well as a load $WR^2$ to be accelerated is treated later in this chapter.

$$W_R = 0.231 \times Wk^2 \times N_s^2 \times (s_i^2 - s_f^2) \times 10^{-6} \tag{3.3}$$

where $W_R$ = secondary $I^2r$ loss, $\text{kW}\cdot\text{sec}$; $Wk^2$ = load + rotor inertia, $\text{lb}\cdot\text{ft}^2$; $N_s$ = synchronous speed, rpm; $s_i$ = initial slip, per unit; and $s_f$ = final slip, per unit.

It is interesting to note that $W_R$ is independent of motor design and depends only on the $Wk^2$ being accelerated and the speed change.

For the case of an acceleration from standstill to full speed, the following simplication can be made

$$s_i^2 - s_f^2 \approx 1.0$$

and the equation for $W_R$ becomes

$$W_R = 0.231 \times Wk^2 \times N_s^2 \times 10^{-6} \tag{3.4}$$

The corresponding $I^2r$ loss in the stator, $W_S$, is given by

$$W_S = W_R \times \left(\frac{r_1}{r_2}\right) \tag{3.5}$$

where $r_1/r_2$ = ratio of the motor's primary resistance to the secondary resistance. It will be noted from the data in Table 3.1 that the $r_1/r_2$ ratio for the Design D motor is much lower than the corresponding value for the Design B motor. This is typical of these two types of designs and is a consequence of the design technique used to achieve the Design D motor's characteristics.

*Design B Motor*

Accelerating losses

$$W_{R_1} = 0.231 \times 50.5 \times 1800^2 \times 10^{-6}$$
$$= 37.8 \text{ kW}\cdot\text{sec}$$

**Table 3.1. Motor Data for Acceleration Calculations**

| | Design B | Design D |
|---|---|---|
| Horsepower | 10 hp | 10 hp |
| Rated load speed | 1770 rpm | 1620 rpm |
| Slip | 1.7% | 10.0% |
| $Wk^2$ of rotor | 0.5 lb·ft$^2$ | 1.0 lb·ft$^2$ |
| $Wk^2$ of load | 50.0 | 50.0 |
| Total $Wk^2$ | 50.5 lb·ft$^2$ | 51.0 lb·ft$^2$ |
| $\frac{\text{Motor primary resistance, } r_1}{\text{Motor secondary resistance, } r_2}$ | 1.50 | 0.36 |
| Efficiency at rated load | 90.2% | 85.4% |

$$W_{S_1} = 1.5 \times 37.8$$
$$= 56.7\ \text{kW}\cdot\text{sec}$$

Running losses

$$W_{F_1} = 10 \times 0.746\left[\frac{1}{0.902} - 1\right]30$$
$$= 24.3\ \text{kW}\cdot\text{sec}$$

$$\text{Total losses for each cycle} = 37.8 + 56.7 + 24.3$$
$$= 118.8\ \text{kW}\cdot\text{sec}$$
$$\text{Ten cycles per hour} \quad \underline{\times 10}$$
$$\text{Total losses per hour of operation} = \underline{1188.0\ \text{kW}\cdot\text{sec}}$$

*Design D Motor*

Accelerating losses

$$W_{R_2} = 0.231 \times 51.0 \times 1800^2(1.0^2 - 0.1^2) \times 10^{-6}$$
$$= 37.8\ \text{kW}\cdot\text{sec}$$
$$W_{S_2} = 0.36 \times 37.8$$
$$= 13.6\ \text{kW}\cdot\text{sec}$$

Running losses

$$W_{F_2} = 10 \times 0.746\left[\frac{1}{0.854} - 1\right]30$$
$$= 38.3\ \text{kW}\cdot\text{sec}$$
$$\text{Total losses for each cycle} = 37.8 + 13.6 + 38.3$$
$$= 89.7\ \text{kW}\cdot\text{sec}$$
$$\text{Ten cycles per hour} \quad \underline{\times 10}$$
$$\text{Total losses per hour of operation} = \underline{897.0\ \text{kW}\cdot\text{sec}}$$

Even though the Design B motor in this example has substantially higher efficiency than the Design D motor, 90.2% versus 85.4%, the total energy consumption for the duty cycle is reduced by applying a Design D motor. As a generalization, if speed changing losses, either acceleration or deceleration losses, comprise a substantial part of the energy consumption, then the use of Design D rather than Design B motors should be evaluated.

## 3.4. Replacement versus Repair

The economics of replacing a burned-out motor with an energy-efficient model versus rewinding the old motor are somewhat problematical because of the uncertainty as to the efficiency which can be achieved by the rewound motor. The efficiency of the failed motor prior to its failure may not be known and the efficiency after rewinding may also be uncertain.

Some rewound motors have been able to duplicate the efficiency of the motor before rewinding, some have evidenced improved efficiency, and others have suffered a loss in efficiency.

One of the major factors involved in determining the success of a rewind is the stripping process used to remove the old winding. Excess heat used in the burnout process can cause irreparable loss in efficiency by degrading the insulation which separates the steel laminations and also can cause mechanical distortion of the stator core and thus air-gap eccentricity when the motor is reassembled. Of course, similar damage may have been caused by localized heating at the time when the motor failed and, if so, no rewinding process will ever fully regain the motor's original efficiency.

This is not to say that all rewinds result in damage to the motor. Some repair shops have addressed the efficiency concern and have established procedures to ensure that the motors are not damaged during stripping and fits and tolerances are acceptable at the time of reassembly. When these practices are adhered to, tests have shown that a motor can be restored to the original efficiency before the motor failed.

The question arises as to whether or not one can expect an efficiency improvement as a result of a rewind. The answer depends to a large degree on the design of the motor prior to rewind and the extent of damage caused by the failure which occasioned the rewind. Recalling the five components of loss discussed in Chapter 2, the one most frequently altered by rewinding is the primary $I^2r$ loss. Increasing the cross-sectional area of conductor material can reduce this loss. Some motors do not have their slots completely full of copper when originally manufactured and, in these cases, an increase in conductor size is possible at the time of rewind. This will effect some reduction in losses and increase in efficiency.

Another technique for loss reduction which is sometimes used involves changing the flux density levels in the machine, either increasing or decreasing the flux densities depending on the original motor design and the requirements of the load and application. This can affect iron losses or secondary $I^2r$ loss. Increasing the flux densities reduces slip and secondary $I^2r$ losses at the expense of core loss and increased inrush current. Reducing the flux densities reduces iron losses but increases secondary $I^2r$ losses. Some test data or design data plus design engineering effort prior to the rewind are required to make a good decision on whether or not a flux change is desirable.

If one assumes that the motor will be restored to its original efficiency, and if this value is known from the manufacturers literature or test, then the economic payback calculations for a new energy-efficient motor versus rewinding can be carried out as described in Chapter 2. The following example illustrates a typical evaluation process:

Example of Repair Versus New Energy-Efficient Motor Economic Evaluation

Hp = 50
Cost differential (CD) repair versus new energy-efficient motor = \$1000.00
Efficiencies: repaired standard motor, $\eta_1 = 90.2\%$
energy-efficient motor, $\eta_2 = 94.1\%$
Operating time per year, $H = 8000$ hr
Cost of electrical energy, $C = \$0.05$ per kWh

$$S = \text{hp} \times 0.746\left[\frac{1}{\eta_1} - \frac{1}{\eta_2}\right]H \times C$$

$$= 50 \times 0.746\left[\frac{1}{0.902} - \frac{1}{0.941}\right] \times 8000 \times 0.05$$

$$= \$685.55$$

$$\text{Payback period, PBP} = \text{CD}/S$$

$$= \frac{1000.00}{685.55}$$

$$= 1.46 \text{ years}$$

The calculation of the payback period in this example shows that the added cost of buying a new, energy-efficient motor versus rewinding the old motor would be recovered in $1\frac{1}{2}$ years.

## 3.5. Multispeed Motors

Multispeed motors can be designed to have torque characteristics equivalent to those shown in Fig. 3.3 for Designs A, B, C, or D. They offer an opportunity for energy savings in situations where the load requirements permit extended periods of reduced speed operation. A good example of this is a ventilating fan which needs to be operated continuously but the fan delivery may be modulated to accommodate needs for high and low airflow rates. For example, a process with a melting furnace may require a high airflow during the work shift when it is in full operation, but only moderate air flow during off hours when the furnace is on standby.

An example will illustrate the energy savings to be realized by applying a multispeed motor to a fan where variable air delivery is desired. Figure 3.4 illustrates the pressure versus volume curve for a

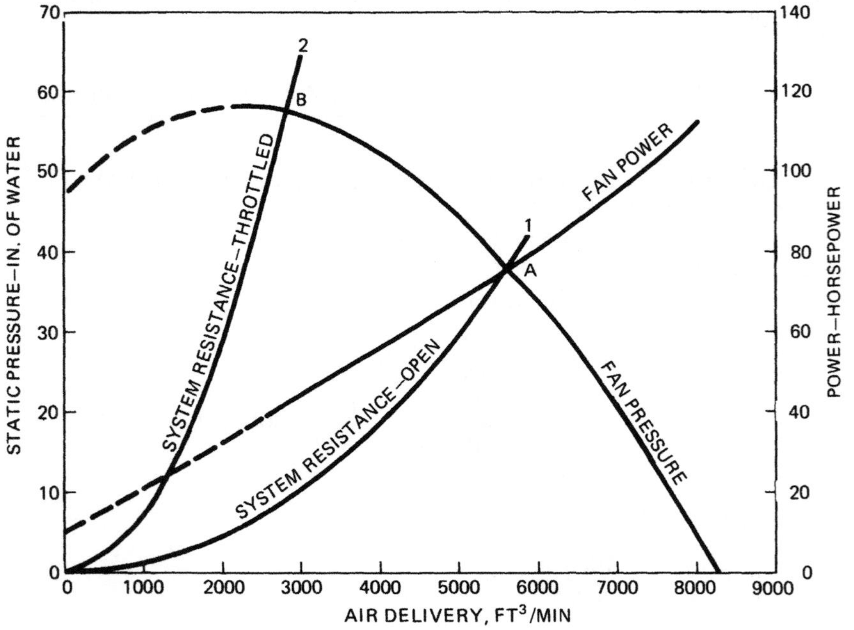


**Figure 3.4.** Characteristic curves for a fan.

typical fan. Also shown are the horsepower versus air delivery line and two system curves superimposed on the fan pressure versus volume curve. The curves can be used in the following manner to determine energy usage. Let curve 1 represent the flow resistance offered to the fan with all of the ducts wide open. The fan will operate at point A on its pressure–volume curve and the input horsepower is 75 hp. Now, to obtain reduced airflow the output is throttled and the new system flow resistance curve is represented by curve 2 which intersects the fan pressure–volume curve at point B. The horsepower required to operate at point B is 43. Assuming an efficiency of 88% for the 75-hp drive motor operating at 43 hp, the electrical input to the motor/fan combination operating at point B is

$$\begin{aligned}\text{Power input} &= \frac{43 \times 0.746}{0.88} \\ &= 36.5 \text{ kW for } 2820 \text{ ft}^3/\text{min air delivery}\end{aligned}$$

An alternate method for controlling airflow is to leave the ventilating system open, no throttling, and reduce the airflow by switching to a low-speed motor connection. A representative speed is half speed resulting in the air delivery being reduced by a factor of 2 and the horsepower required to drive the fan being reduced by $(1/2)^3 = 1/8$ of the rated 75 hp. A multispeed motor is usually less efficient than equivalent single-speed motors of the same horsepower rating. Assume the efficiency for the multispeed motor to be 80% on low speed at the required horsepower. The power required for the same ft$^3$/min of air delivery as before is

$$\begin{aligned}\text{Power input} &= \frac{9.375 \times 0.746}{0.80} \\ &= 8.7 \text{ kW for } 2820 \text{ ft}^3/\text{min air delivery}\end{aligned}$$

The energy savings is 27.8 kW in favor of the multispeed motor versus throttling the output of a constant-speed motor drive.

Multispeed motors are available in a multitude of varieties. Equation (3.1) gave the relationship between the synchronous speed of an ac machine and the number of poles for which the machine is wound. The most frequently used multispeed arrangement provides

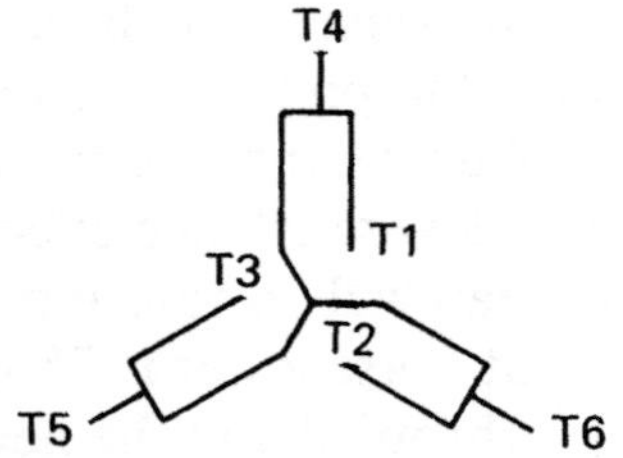


| Speed | L1 | L2 | L3 | Open | Join |
|---|---|---|---|---|---|
| Low | T1 | T2 | T3 | T4,T5,T6 | — |
| High | T6 | T4 | T5 | — | T1,T2,T3 |

**Figure 3.5.** Variable-torque two-speed single winding motor.

two speeds differing by 2:1 with the speed change being accomplished by a change in external connections. This type of motor is a single-winding, multispeed machine and there are three types of designs in common usage. The connection diagrams for the various types of multispeeds have been published by Veinott[2] and are reproduced in Figs. 3.5 through 3.11.

1. *Variable torque:* The breakdown and full-load torques vary directly as the speed. Therefore, since both the lower speed and torques associated with it are 50% of the corresponding high-speed values, the low-speed horsepower is 25% of the high-speed horsepower. This characteristic is particularly useful for fans and centrifugal pumps since their horsepower varies as the cube of the speed. The connection diagram for a variable-torque, two-speed, single-winding motor is shown in Fig. 3.5.

   Physically, inside the motor the change from the parallel

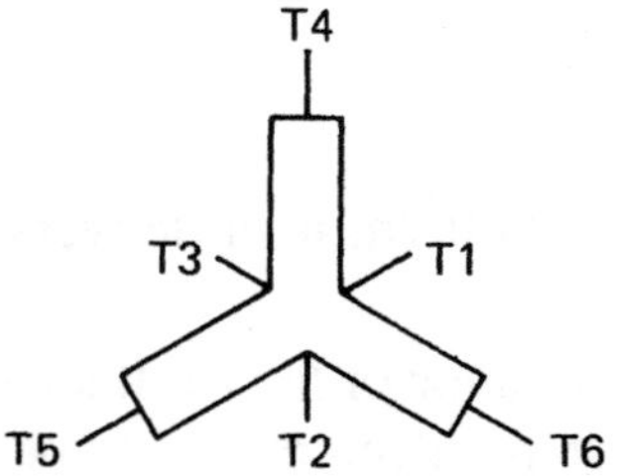


| Speed | L1 | L2 | L3 | Open | Join |
|---|---|---|---|---|---|
| Low | T1 | T2 | T3 | T4,T5,T6 | — |
| High | T6 | T4 | T5 | — | T1,T2,T3 |

**Figure 3.6.** Constant-torque two-speed single winding motor.

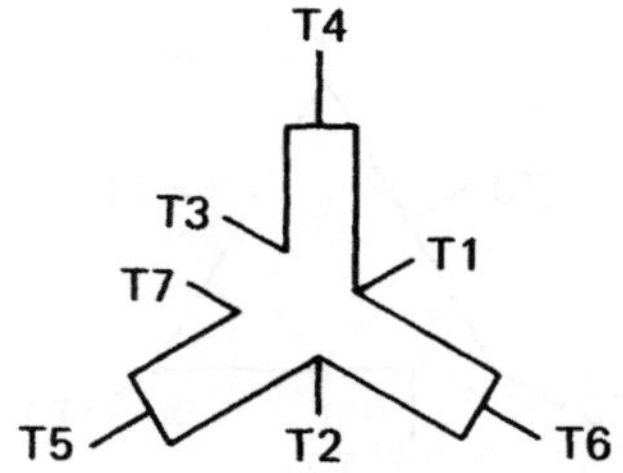


| Speed | *L*1 | *L*2 | *L*3 | Open | Join |
|---|---|---|---|---|---|
| Low | *T*1 | *T*2 | *T*3,*T*7 | *T*4,*T*5,*T*6 | —— |
| High | *T*6 | *T*4 | *T*5 | | *T*1,*T*2,*T*3,*T*7 |

**Figure 3.7.** Constant-torque two-speed single winding multispeed motor with two or more independent windings.

Y connection for high speed to a series Y for the low speed reverses the direction of the current in half of the pole groups in the winding. If the machine has been wound for four poles in high speed, the poles around the periphery of the air gap are excited north, south, north, and south at some particular instant of time. In the low-speed connection, due to the reversal of half of the pole groups, the excited poles would be north, north, north, and north. The intervening space would then become magnetized as a "consequence" to provide an effective north, south, north, south, north, south, north, and south, eight-pole pattern. The low-speed connection of a single-winding multispeed is often referred to as the "conse-

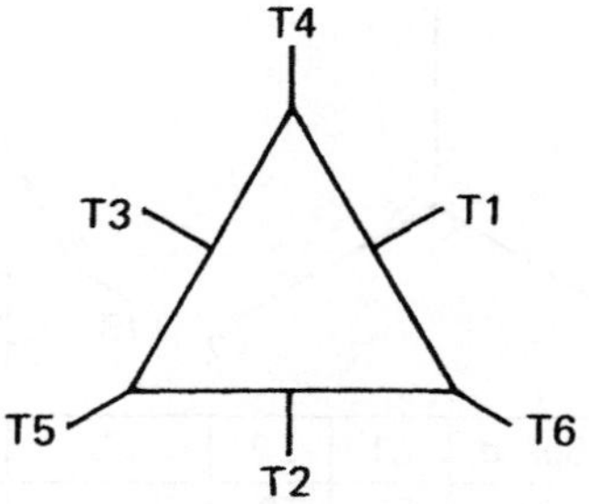


| Speed | *L*1 | *L*2 | *L*3 | Open | Join |
|---|---|---|---|---|---|
| Low | *T*1 | *T*2 | *T*3 | —— | *T*4,*T*5,*T*6 |
| High | *T*6 | *T*4 | *T*5 | *T*1,*T*2,*T*3 | —— |

**Figure 3.8.** Constant-horsepower two-speed single-winding motor.

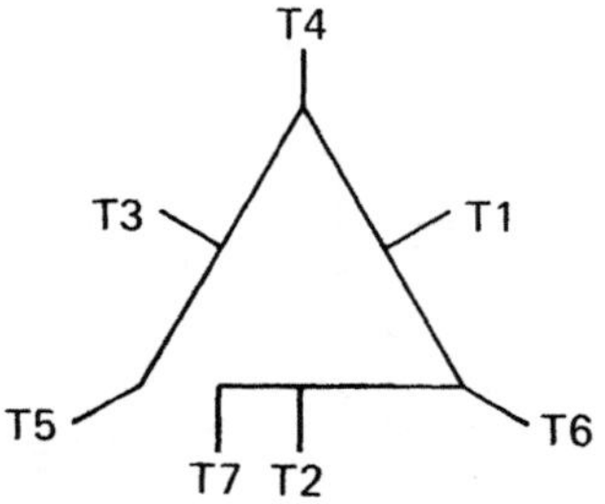


| Speed | L1 | L2 | L3 | Open | Join |
|---|---|---|---|---|---|
| Low | T1 | T2 | T3 | —— | T4,T5,T6,T7 |
| High | T6 | T4 | T5,T7 | T1,T2,T3 | |

**Figure 3.9.** Constant-horsepower two-speed single winding multispeed motor with two or more independent windings.

quent pole" connection. All of the single-winding, multispeed arrangements illustrated in Figs. 3.5–3.9 utilize this same principle to achieve the two speeds.

2. *Constant torque:* Both the breakdown torque and full-load torque remain approximately constant with this connection. The low-speed horsepower is, therefore, 50% of the high-speed value. The connection on high speed is a parallel Y, and for the low speed a series delta is used as shown in Fig. 3.6.

   If a second winding is present in the machine, the delta should be left open when not in use to avoid the possibility of

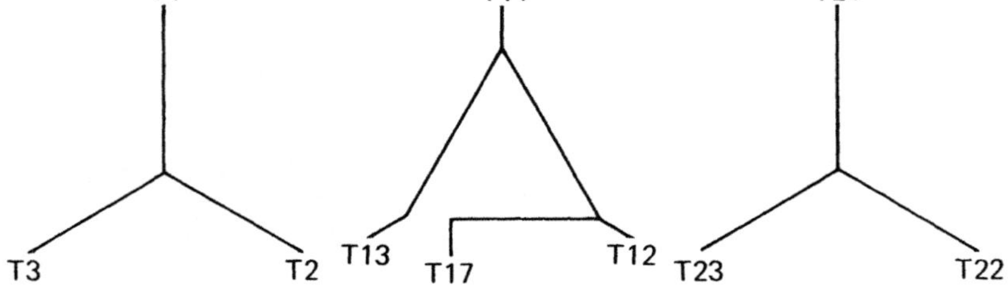


| Speed | L1 | L2 | L3 | Open | Join |
|---|---|---|---|---|---|
| Low | T1 | T2 | T3 | T11,T12,T13,T17,T21,T22,T23 | — |
| 2nd | T11 | T12 | T13,T17 | T1,T2,T3,T21,T22,T23 | — |
| High | T21 | T22 | T23 | T1,T2,T3,T11,T12,T13,T17 | — |

**Figure 3.10.** Three-speed three-winding motor.

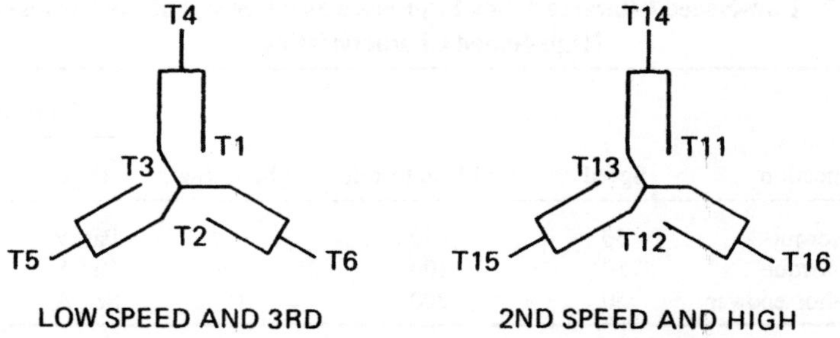


| Speed | L1 | L2 | L3 | Open | Join |
|---|---|---|---|---|---|
| Low | *T1* | *T2* | *T3* | *T4,T5,T6* | — |
| 2nd | *T11* | *T12* | *T13* | *T14,T15,T16* | — |
| 3rd | *T6* | *T4* | *T5* | — | *T1,T2,T3* |
| High | *T16* | *T14* | *T15* | — | *T11,T12,T13* |

**Figure 3.11.** Four-speed two-winding motor, each variable torque.

circulating currents being induced from the other winding. The connection to be used in this case is shown in Fig. 3.7.

Typical applications for the constant-torque connection are conveyors, positive displacement pumps, and compressors.

3. *Constant horsepower:* The breakdown and full-load torques vary inversely proportional to speed so that the horsepower remains constant. The high-speed connection is series delta and the low-speed connection is parallel Y as shown in Fig. 3.8. Figure 3.9 shows the alternate connection scheme to open the delta if a second winding is excited. Constant-horsepower motors are used on machine tools, winches, and similar applications.

Table 3.2 summarizes the three types of single-winding multispeed motors. Additional types of single-winding multispeed motors developed by Rawcliffe[3–5] and others are in use to achieve speed ratios other than 2:1. Some of the pole combinations available using Rawcliffe's windings are 4/6, 6/8, 8/10, and 10/12. These arrangements have been termed "pole amplitude modulation" windings because they utilize a principle similar to the amplitude modulation technique used for AM radio transmission. For one of the two speeds the pole phase groups are arranged similar to a conventional

**Table 3.2. Low-Speed Characteristics Expressed as a Percent of the Corresponding High-Speed Characteristics[a]**

| | | | | Connection | |
|---|---|---|---|---|---|
| Connection | Speed | Full-load torque | hp rating | High | Low |
| Variable-torque | 50 | 50 | 25 | Par. Y | Ser. Y |
| Constant-torque | 50 | 100 | 50 | Par. Y | Ser. Δ |
| Constant-horsepower | 50 | 200 | 100 | Ser. Δ | Par. Y |

[a] From Veinott.[2]

polyphase winding, for example an 8-pole winding. The second speed is achieved by reconnecting some of the pole phase groups to achieve the effect of having the 8-pole magnetic field modulated by another field of 2 poles. The results of amplitude modulation are two new frequencies that are the sum and difference of the original frequencies which, for the example case, are 6- and 10-pole fields. The reconnection results in suppression of the 6-pole field and the 10-pole field dominates to produce the second speed. Several selected references among the many technical papers written on this subject are given at the end of this chapter.

Considerable flexibility in speed range can be achieved by using two or more windings in the motor. Each winding can be designed to provide two speeds using one of the connection arrangements from Table 3.2.

Theoretically a very large number of possible speed combinations could be achieved in this manner. However, as in most design situations, there are some practical limits. The magnetic structure of a multispeed motor is not utilized with equal effectiveness for all speeds. The design compromises necessary in multispeed motors mean that they are usually physically larger than the single-speed motor of the equivalent horsepower rating. The most often used speed combinations are:

- *Two speed, two winding.* Any even number of poles and any combination of horsepower ratings are possible.
- *Three speed, three winding.* Any even number of poles and any combination of horsepower ratings are possible. Magnetic material utilization is almost certainly going to be inefficient on one or more of the windings and consequently the machine

size will likely be the limiting factor in this arrangement. A typical connection diagram is shown in Fig. 3.10.

- *Three speed, two winding.* One winding is used to provide two speeds in accordance with Table 3.2. The second winding can provide an independent selection of speed and horsepower.
- *Four speed, two winding.* In this case both windings are reconnectable for two speeds giving a total of four speeds. Figure 3.11 shows two variable torque connections but any combination of the winding arrangements from Table 3.2 can be used.

In multispeed motors with more than one winding, delta connections should be opened when the delta winding is idle and another winding in the machine is excited. If the delta is not opened, the possibility of circulating currents induced from the active winding exists and this can result in excess motor heating.

## 3.6. Load Shedding Saves Energy

Load shedding refers to the practice of turning a motor off when it is not needed instead of operating the motor at no load. Energy savings can be achieved through this practice provided the energy consumed by restarting does not exceed the energy saved during the off period.

The energy savings by load shedding can be evaluated with the aid of Eqs. (3.4) and (3.5). One also must know the motor's losses under no load. The following example will illustrate the evaluation process.

EXAMPLE: LOAD SHEDDING CALCULATION. A 10-hp, Design B motor is driving a conveyor. Design data for the motor are given in Table 3.1. When the conveyor is empty, the input to the motor is 2.0 kW. The process loads the conveyor and requires it to operate only 5 minutes out of every 15. Should the motor be turned off for the 10 minutes that it is not needed, or should the motor and conveyor be left operating continuously?

The conveyor $WR^2$ is 2.0 lb · ft$^2$ and the motor control is arranged so that the motor can be started and reach full speed just prior to the arrival of the load.

$$W_R = 0.231 \times (2.0 + 0.5) \times 1800^2 \times 10^{-6}$$
$$= 1.871 \text{ kW} \cdot \text{sec}$$
$$W_S = 1.871 \times 1.5$$
$$= 2.807 \text{ kW} \cdot \text{sec}$$

$$\text{Accelerating losses} = 1.871 + 2.807$$
$$= 4.678 \text{ kW} \cdot \text{sec per start}$$

$$\text{Energy consumed in 1 sec of no-load running} = 2.0 \text{ kW} \cdot \text{sec}$$

$$\text{Running time at no load to equal 1 start} = 4.678/2.0$$
$$= 2.339 \text{ seconds}$$

Therefore, if the motor is idle longer than 3 seconds, it is economical to turn the motor off. Since the idle time is 10 minutes out of every 15, clearly the motor should be shut off. The energy savings per 15-minute cycle saved by shutting the motor off are

$$\text{Energy savings} = (2.0 \times 10 \times 60 - 4.678)/(60 \times 60)$$
$$= 0.332 \text{ kWh per 15-minute cycle}$$

Another example illustrates the advantages of load shedding in multimotor installations.

EXAMPLE: LOAD SHEDDING IN A MULTIMOTOR INSTALLATION. Five 100-hp, four pole motors are used to supply the comfort and process cooling requirements for a manufacturing plant. The plant operates 16 hours per day, 5 days per week, 50 weeks per year. For 6 hours per day 100% of the cooling capacity is required. For the remaining 10 hours of the work shift, only 60% capacity is required and for all of the nonworking time, 20% capacity is required. What are the savings to be realized by shutting the motors off as opposed to operating them at no load when they are not required? The no-load losses of these 100-hp motors are 13.0 kW and the full-load input is 80.0 kW. The rotor plus load $WR^2$ for each motor is 40 lb·ft$^2$ and the motors are started unloaded. The optimum matching of the motor horsepower with the plant load requirements requires an average number of starts per motor of 4 during the 16-hour work shift.

Energy Usage per Year—Motors Operating Continuously

| | days/yr | | hr/day | | No. of motors | | kW | | |
|---|---|---|---|---|---|---|---|---|---|
| 100% load | 250 | × | 6 | × | 5 | × | 80 | = | $600 \times 10^3$ kWh |
| 60% load | 250 | × | 10 | × | 3 | × | 80 | = | $600 \times 10^3$ |
| | 250 | × | 10 | × | 2 | × | 13 | = | $65 \times 10^3$ |
| 20% load | 250 | × | 8 | × | 1 | × | 80 | = | $160 \times 10^3$ |
| | 250 | × | 8 | × | 4 | × | 13 | = | $104 \times 10^3$ |
| | 115 | × | 24 | × | 1 | × | 80 | = | $220.8 \times 10^3$ |
| | 115 | × | 24 | × | 4 | × | 13 | = | $143.5 \times 10^3$ |
| | | | | | Total energy usage per year | | | = | $1893.3 \times 10^3$ kWh |

Energy Usage per Year—Motors Cycling On and Off

| | days/yr | | hr/day | | No. of motors | | kW | | |
|---|---|---|---|---|---|---|---|---|---|
| 100% load | 250 | × | 6 | × | 5 | × | 80 | = | $600 \times 10^3$ kWh |
| 60% load | 250 | × | 10 | × | 3 | × | 80 | = | $600 \times 10^3$ |
| 20% load | 250 | × | 8 | × | 1 | × | 80 | = | $160 \times 10^3$ |
| | 115 | × | 24 | × | 1 | × | 80 | = | $220.8 \times 10^3$ |
| | | | | | Total for loaded periods | | | = | $1580.8 \times 10^3$ kWh |

Starting energy usage

$$W_R = 0.231 \times 40 \times 1800^2 \times 10^{-6}$$

$$= 29.938 \text{ kW}\cdot\text{sec}$$

$$W_S = 29.938 \times 1.5$$

$$= 44.907 \text{ kW}\cdot\text{sec}$$

$$\text{Accelerating losses} = 29.938 + 44.907$$

$$= 74.845 \text{ kW}\cdot\text{sec per motor start}$$

$$\text{Starting energy usage} = \frac{74.845 \times 250 \times 4 \times 5}{3600}$$

$$= 104.0 \text{ kWh}$$

$$\text{Total energy usage per year} = (1580.8 + 0.1)10^3$$

$$= 1580.9 \times 10^3 \text{ kWh}$$

The net savings achieved by cycling the motors on and off are

$$\text{Energy savings} = (1893.3 - 1580.9) \times 10^3$$
$$= 312.4 \times 10^3 \text{ kWh}$$

Assuming a cost of \$0.05 per kWh for electricity, the cost savings are

$$\text{Cost savings} = 312.4 \times 10^3 \times 0.05$$
$$= \$15{,}620 \text{ per year}$$

The calculations in the preceding examples are based on comparing the energy consumed during an acceleration with the energy that would be expended if the motor was operated at no load when not needed instead of being turned off. This type of calculation determines if load shedding should be considered. Once it is decided that load cycling will yield energy savings, then the questions of the motor and power supply system's capability to operate under cycling conditions must be addressed.

Voltage reductions caused by starting motors can be troublesome. Computers and other electronic equipment are often particularly sensitive to voltage dips caused by motor starting. A separate power supply for the voltage-sensitive equipment provides a means of isolating this equipment from the effects of motor cycling.

The question of how frequently a motor can be started is one which often requires individualized analysis for each application. Thermal, mechanical, and winding stress are all factors which influence the permissible number of starts. When a motor is started, both stator and rotor conductors carry currents much larger than they do under normal operating conditions. Typically, during the initial inrush which occurs during starting, the rms value of current will equal approximately six times the rated load value which, for this brief period, results in $I^2r$ losses in excess of 36 times the $I^2r$ losses generated during full-load operation. The current and $I^2r$ losses associated with this current decline as the motor accelerates, but over most of the accelerating period the current and losses remain several times larger than at normal operating speed and rated load. Because of the relatively short time during which these losses are generated, the motor cannot dissipate the losses to the surrounding iron and outside air, and, consequently, much of the energy is stored in the conductors where it is generated. This causes the temperature of both stator and rotor conductors to rise rapidly.

Based on the assumption that all of the $I^2r$ losses occurring during one acceleration are stored in the conductor material, the temperature rise of the conductor can be calculated from the equation

$$°\text{C temperature rise} = \frac{\text{kW} \cdot \text{sec per start} \times 10^3}{wC} \tag{3.6}$$

where $w$ = weight of conductor material in lb and $C$ = thermal capacity of material in W · sec/lb · °C; for copper $C = 176$ and for aluminum $C = 430$. Of course, the all-heat-stored assumption is conservative and the temperature rise actually experienced will be somewhat less. However, Eq. (3.6) is widely used and gives good results for accelerations lasting only a few seconds. For longer accelerating periods, 15 seconds or more, some modification needs to be introduced to account for the heat dissipated to the surrounding iron and outside air.

A better understanding of the heating that occurs during acceleration can be obtained from the following typical example:

A 300-hp, 1200-rpm motor is driving a ventilating fan. The total motor plus connected $Wk^2$ is 6000 lb · ft² which is approximately twice the NEMA standardized value for this size of motor. This is not unusual for a ventilating fan drive in this size range. Other pertinent data are:

Kilowatt seconds for one acceleration, stator = 2800 kW · sec

Kilowatt seconds for one acceleration, rotor = 2500 kW · sec

Time for one acceleration = 17.8 sec

Weight of stator copper winding = 250 lb

Weight of rotor aluminum cage = 100 lb

The motor must provide torque to both accelerate the $Wk^2$ and deliver the torque load required by the fan. A method for calculating the kilowatt seconds dissipated as motor losses during this type of acceleration is given in Section 3.7.

Using Eq. (3.6), the temperature rise of the stator and rotor conductors are:

$$\text{Stator winding temperature rise} = \frac{2800 \times 10^3}{176 \times 250}$$

$$= 63.6°\text{C}$$

$$\text{Rotor cage temperature rise} = \frac{2500 \times 10^3}{430 \times 100}$$

$$= 58.1°\text{C}$$

A single acceleration does not elevate the temperatures to damaging values, but several successive accelerations clearly would.

It is evident from the preceding example that motor heating can be a problem when loads are cycled on and off. Frequent starts which cause the stator windings to exceed their allowable temperature rise, will eventually degrade insulation and cause premature motor failure.

In some cases, rotor cage temperatures can also reach values which will shorten motor life. Cast aluminum rotors begin to soften at temperatures exceeding 250°C, although aluminum does not melt until the temperature reaches approximately 600°C. In larger horsepower ratings, the cage is constructed of individual bars inserted through the rotor slots and joined to the end ring by brazing or welding. This type of cage construction is more vulnerable to temperature cycling than cast rotors. Each time the temperature is elevated, both the bars and end rings attempt to expand, giving rise to stress in the bars. Alternate heating and cooling of the rotor cage can cause eventual fatigue failure of the rotor.

Stator winding temperature protectors are often used in large machines to protect against excess stator temperatures. These temperature detectors take the form of resistance temperature detectors or thermocouples mounted in the stator winding or thermostats mounted on the coil heads. It should not be assumed that this type of protection will necessarily protect the motor against cycling. If the motor is rotor limited, that is, the rotor reaches dangerous temperatures before the stator, then a winding temperature protector will not save the motor from failure caused by too many starts and stops. If cycling duty is anticipated, it should be stated at the time the motor is purchased so that there is no misunderstanding the role of the temperature protecting devices.

The large currents which occur during across-the-line starting cause some movement of the stator winding coil heads. These stress cycles on the coil heads can eventually cause failure if the coils are not properly braced to withstand repeated starts.

Another cost associated with load cycling is life reduction in the motor starter. Electromagnetic contractors have a limited number of starts before the contacts begin to wear or pit, linkages fail, and solenoid coils burn out. Nailen[9] cites a 1974 survey of plant engineers which revealed that on systems of 600 volts and below, motor starter failures were a third more frequent than motor failures. Therefore, the starter must also be selected to provide acceptable life for a load cycling application.

There are alternatives which can be used instead of cycling motors on and off during off-peak load periods:

1. Inlet guide vane control can be used for fans to reduce airflow and motor load during off-peak requirements. This technique can be used over a limited flow range and is more efficient than output throttling.
2. Use a clutch to disengage the motor from the load.

Energy savings which can be achieved by cycling loads are too attractive not to be used. However, one needs to be aware of some of the considerations cited above so that the motor and its control equipment are properly applied, thereby avoiding some of the hidden costs of cycling which might offset the advantages gained.

A guideline for use in load shedding applications which is based on a fan or pump load and a $Wk^2$ equal to the motor inertia is a maximum of four starts per hour allowable up to 50 hp, three starts per hour above 50 hp up to 100 hp, and specific recommendations from the motor manufacturer above 100 hp. For two-pole motors only, the allowable number of starts is three up to 50 hp and two from 50 to 100 hp. This rule is conservative and may be modified as specific data are available for the motor being used.

Computers are being used to implement load cycling schedules. A process control computer incorporating digital and analog data interfaces can be used to control the cycling of motors and other energy-consuming devices based on time of day, time duration, temperature, pressure, flow, or many other types of process conditions. Some control systems turn devices on and off based on time of day and day of the week. More sophisticated controls sense conditions from the area being controlled such as temperature and humidity and then combine this information with the time of day to control the environment to different limits depending on whether people are working in the area or not. Of course, computer control is not limited to environmental control for people. Processes which are computer directed often incorporate control algorithms which seek to minimize the energy expenditure. The design of these systems should always include an analysis of the motor cycling requirements to ensure that they comply with the principles outlined in this section.

## 3.7. Calculation of Acceleration Losses when Retarding Load Torque is Present

Equations (3.3), (3.4), and (3.5) have been used in the examples so far in this chapter to calculate motor losses during acceleration. As stated in the text accompanying Eq. (3.3), it was assumed that there was no appreciable retarding load torque present during the acceleration. This is representative of many classes of applications. A pump which is unloaded during the starting period is a typical example. However, there are other applications which do impose a load on the motor in addition to the $Wk^2$ which is being accelerated. Notable among these types of loads are fans for which the load torque varies as the square of the speed. Whenever the load torque is significant, (3.3) does not apply and another calculation method must be used.

Figure 3.12 displays motor torque and a representative fan load-

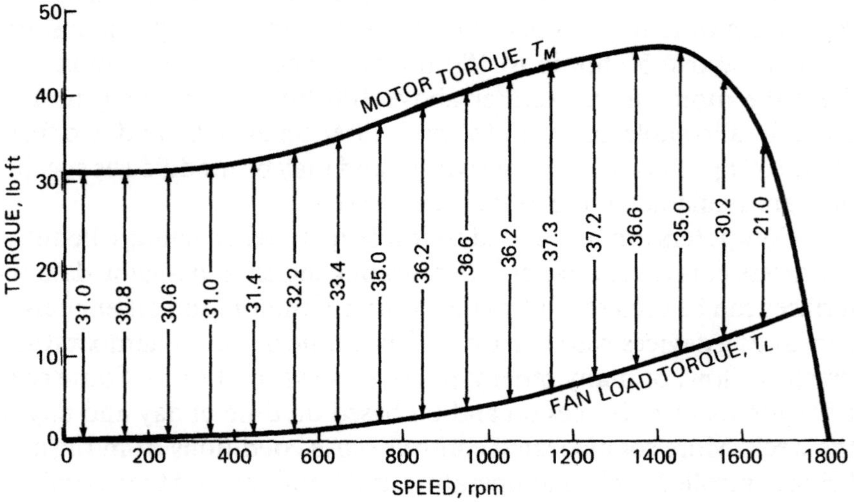


**Figure 3.12.** Motor and fan load torque versus speed.

torque curve, both plotted against speed. The underlying assumption for the relationship in (3.3) to be valid is that all of the motor torque is available to accelerate the motor inertia. In Fig. 3.12 only the difference between the motor torque and the load torque, $T_M - T_L$, is available for acceleration.

The procedure for calculating acceleration losses in this case is to divide the speed–torque curve into segments small enough that the torques, both motor and load torques, can be assumed to be constant at some average value over the speed interval. The division of the curves into segments is shown in Fig. 3.12. The secondary loss for each speed segment can be calculated from the equation

$$W_{RI} = 0.231 \times Wk^2 \times N_s^2 \times \frac{T_{MI}}{T_{MI} - T_{LI}} [s_{iI}^2 - s_{fI}^2] \times 10^{-6} \quad (3.7)$$

where $W_{RI}$ = secondary loss for the $I$ speed interval, kW·sec; $T_{MI}$ = average motor torque for the $I$ speed interval, lb·ft; $T_{LI}$ = average load torque for the $I$ speed interval, lb·ft; $s_{iI}$ = slip at beginning of $I$ speed interval, per unit; $s_{fI}$ = slip at end of $I$ speed interval, per unit; and other terms were defined with Eq. (3.3).

The stator loss for the $I$ speed increment is calculated using

$$W_{SI} = W_{RI}\left(\frac{r_{1I}}{r_{2I}}\right) \quad (3.8)$$

The subscript $I$ is included with the $r_1$ and $r_2$ terms to denote the fact that both of these terms may vary during the accelerating period. The variation in $r_1$ is related to temperature change resulting from the accumulation of heat in the stator winding during the acceleration. The value of $r_2$ also changes because of temperature but, in addition, the change in the slip frequency of current in the rotor changes the effective rotor resistance. If the rotor bars are deep enough to evidence a current crowding, i.e., skin effect, during stalled rotor, where the rotor frequency is equal to the line frequency, then the rotor resistance will decline as the rotor speed increases. This phenomenon is described by Alger[6] and others and is present in most induction motors of sizes approximately 1 hp and above.

The procedure for calculating accelerating losses is illustrated in

**Table 3.3. Calculation of Acceleration Losses for 5-hp Motor Accelerating a Fan Load**

Motor data: 5 hp, 1750 rpm
460 V, 3 phase, 60 Hz
$Wk^2$ (load + motor rotor) = 1.2 lb·ft$^2$
$r_1/r_2$ = 1.1 assumed constant at average value over speed range

| Speed rpm | $T_{MI}$ lb·ft | $(T_{MI} - T_{LI})$ lb·ft | $s_{iI}$ P.U. | $s_{fI}$ P.U. | $W_{RI}$ W·sec | $W_{SI}$ W·sec | $W_{RI} + W_{SI}$ W·sec |
|---|---|---|---|---|---|---|---|
| 0–100 | 31.0 | 31.0 | 1.000 | 0.944 | 97.0 | 106.7 | 203.7 |
| 100–200 | 31.0 | 30.8 | 0.944 | 0.889 | 92.1 | 101.3 | 193.4 |
| 200–300 | 31.2 | 30.6 | 0.889 | 0.833 | 87.6 | 96.4 | 184.0 |
| 300–400 | 31.8 | 31.0 | 0.833 | 0.778 | 82.5 | 90.7 | 173.2 |
| 400–500 | 32.3 | 31.4 | 0.778 | 0.722 | 77.0 | 84.7 | 161.7 |
| 500–600 | 33.3 | 32.2 | 0.722 | 0.667 | 71.7 | 78.8 | 150.5 |
| 600–700 | 35.0 | 33.4 | 0.667 | 0.611 | 66.8 | 73.5 | 140.3 |
| 700–800 | 37.0 | 35.0 | 0.611 | 0.555 | 61.5 | 67.7 | 129.2 |
| 800–900 | 39.0 | 36.2 | 0.555 | 0.500 | 56.7 | 62.4 | 119.1 |
| 900–1000 | 40.5 | 36.6 | 0.500 | 0.444 | 52.1 | 57.4 | 109.5 |
| 1000–1100 | 42.0 | 36.2 | 0.444 | 0.389 | 48.2 | 53.1 | 101.3 |
| 1100–1200 | 43.5 | 37.3 | 0.389 | 0.333 | 42.0 | 46.2 | 88.2 |
| 1200–1300 | 44.2 | 37.2 | 0.333 | 0.278 | 36.2 | 39.9 | 76.1 |
| 1300–1400 | 45.0 | 36.6 | 0.278 | 0.222 | 30.7 | 33.7 | 64.4 |
| 1400–1500 | 45.0 | 35.0 | 0.222 | 0.167 | 24.9 | 27.4 | 52.3 |
| 1500–1600 | 42.0 | 30.2 | 0.167 | 0.111 | 19.3 | 21.2 | 40.5 |
| 1600–1700 | 32.1 | 21.0 | 0.111 | 0.056 | 12.7 | 14.0 | 26.7 |
| 1700–1750 | 20.1 | 5.4 | 0.056 | 0.028 | 7.7 | 8.5 | 16.2 |
| Total | | | | | 966.7 | 1063.7 | 2030.4 |

Fig. 3.12 and Table 3.3 in which the loss calculation for a 5-hp motor accelerating a fan load is detailed.

## 3.8. Unbalanced Voltage

Operation on unbalanced voltages is detrimental not only to motor efficiency but to torques and heating as well. Voltage unbalance usually results from a nonuniform distribution of single-phase loads connected to a three-phase system. NEMA specifies a "derating factor" to be applied to the motor's nominal horsepower rating based on the percent of voltage unbalance. The NEMA derating curve is

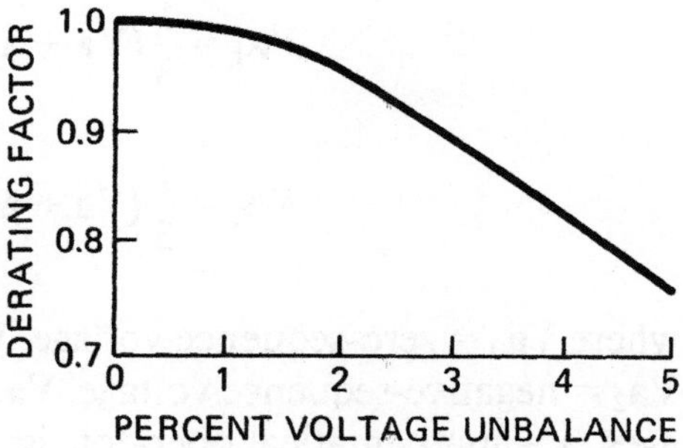


**Figure 3.13.** Motor derating factor related to unbalance voltage. (This is Fig. 14-1 of NEMA Standards Publication MG 1-1993. It is reproduced by permission of the National Electrical Manufacturers Association which holds the copyright.)

displayed in Fig. 3.13. Voltage unbalance, the abscissa of Fig. 3.13, is calculated from the equation*

Percent voltage unbalance

$$= 100 \times \frac{\text{Maximum voltage deviation from average voltage}}{\text{Average voltage}} \tag{3.9}$$

Example: With voltages of 460, 467, and 450, the average is 459, the maximum deviation from the average is 9, and the percent unbalance = 100 × 9/459 = 1.96%.

Consider a voltage unbalance of 3%. From Fig. 3.13, the derating factor is approximately 0.9, which viewed as a percentage is far in excess of the voltage unbalance percentage. A small percentage of voltage unbalance results in a large increase in current over the value needed for the same load with balanced voltages. To explain this phenomenon, it is convenient to introduce the concept of positive-, negative-, and zero-sequence voltages. A means for analyzing motor operation under unbalanced voltage conditions is to first resolve the unbalanced voltages into their sequence components. Clarke,[7] Wagner and Evans,[8] and others give the equations for calculating the three sequence voltages

$$\mathbf{Va_0} = \frac{1}{3}(\mathbf{Va} + \mathbf{Vb} + \mathbf{Vc}) \tag{3.10}$$

* The equation and example are 14.35.2 of NEMA Standards Publication MG1-1993. It is reproduced by permission of the National Electrical Manufacturers Association which holds the copyright.

$$\mathbf{Va}_1 = \frac{1}{3}(\mathbf{Va} + \mathbf{aVb} + \mathbf{a}^2\mathbf{Vc}) \tag{3.11}$$

$$\mathbf{Va}_2 = \frac{1}{3}(\mathbf{Va} + \mathbf{a}^2\mathbf{Vb} + \mathbf{aVc}) \tag{3.12}$$

where $\mathbf{Va}_0$ = zero-sequence voltage, $\mathbf{Va}_1$ = positive-sequence voltage, $\mathbf{Va}_2$ = negative-sequence voltage, **Va, Vb, Vc** = a, b, and c phase voltages, a = mathematical operator signifying rotation of the phasor on which it operates by 120°, and $a^2$ = operator signifying rotation by 240°. All of the voltages in (3.10)–(3.12) are phasor quantities and they may be either line-to-line or line-to-neutral quantities. For the discussion which follows, **Va, Vb,** and **Vc** will be considered to be phase voltages rather than line-to-line voltages because the induction motor equivalent circuit will be utilized and this equivalent circuit is a circuit representing one phase of a polyphase machine.

The effect of performing the transformation given in (3.10) through (3.12) is to convert the original unbalanced set of voltages into three sets of balanced voltages, one having a positive-sequence rotation, i.e., same direction as the original voltages, one having a negative-sequence, i.e., opposite direction, and the third, zero-sequence, voltages all in phase with one another. The relationship of these phasor quantities is illustrated in Fig. 3.14.

It is known that zero-sequence voltages resulting from an unbalanced set of line voltages will not cause any zero-sequence current in the motor windings so long as the motor neutral is not connected to ground or any other return path to the power supply. Since this is the case in the vast majority of motor installations, no further consider-

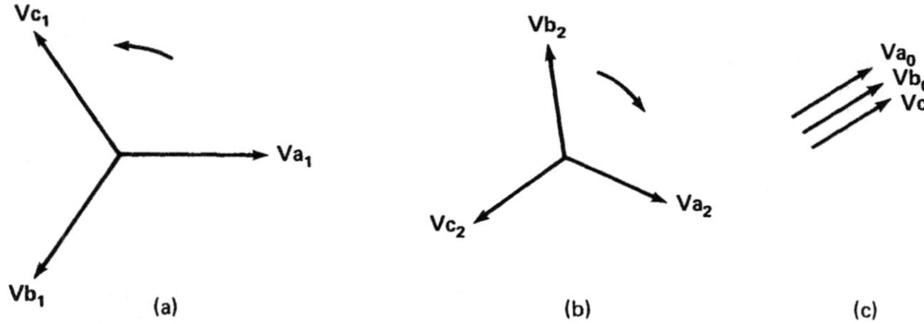


**Figure 3.14.** (a) Positive-, (b) negative-, and (c) zero-sequence phasors.

ation will be given to the zero-sequence component. The effects of the positive- and negative-sequence voltages can be determined by using the principle of superposition. The positive-sequence voltage causes the magnetic field in the motor to rotate in the forward direction and thus produces a torque in that direction. The motor will, in fact, rotate in the same direction as it would with balanced voltages applied for most voltage unbalances experienced in practical situations. The negative-sequence voltage will cause a negatively rotating field and consequently a negative torque so that the motor's net torque will be reduced as a result of the voltage unbalance. The motor can be represented by two equivalent circuits, one for the positive-sequence and one for the negative-sequence voltages as shown in Fig. 3.15. The slip for the positive-sequence circuit is the same as usually calculated for an induction motor,

$$s_p = \frac{n_s - n}{n_s} \tag{3.13}$$

However, the equation for slip of the negative-sequence circuit is different because of the opposite direction of rotation and is given by

$$s_n = \frac{n_s + n}{n_s} \tag{3.14}$$

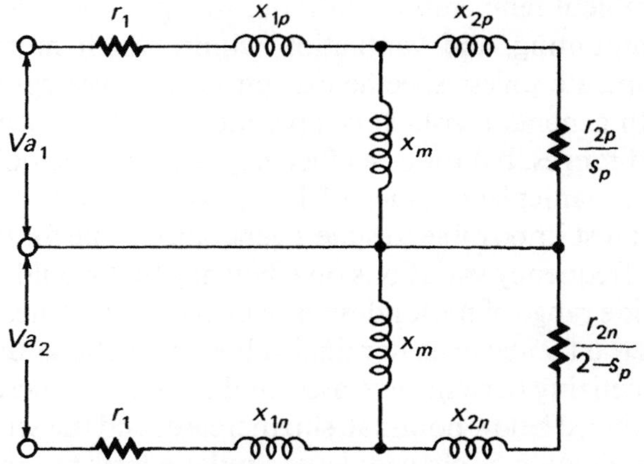


Impedance subscripts:
*p* designates positive-sequence impedance
*n* designates negative-sequence impedance

**Figure 3.15.** Induction motor equivalent circuit for unbalanced voltage operation.

where $s_p$ and $s_n$ = slips for positive- and negative-sequence circuits, respectively, $n_s$ = synchronous speed, and $n$ = actual speed. Combining (3.13) and (3.14), the slip for the negative-sequence circuit is

$$s_n = 2 - s_p \tag{3.15}$$

The equivalent circuit representation for an induction motor operating on unbalanced voltages is displayed in Fig. 3.15.

Examination of the negative-sequence circuit reveals why small unbalances in voltage cause relatively large negative-sequence currents. Assume the motor is running at a slip of 0.02. Then $2 - s_p = 1.98$ and the $r_2/s_n$ term in the negative-sequence circuit is $r_2/1.98$ which makes the motor's impedance to negative-sequence voltages appear very low, generally somewhat lower than the motor's impedance at locked rotor.

These negative-sequence currents result in $I^2r$ losses and negative torque both of which are wasted energy and should be avoided.

## 3.9. Other Unusual Service Conditions

NEMA[1] tabulates a list of usual service conditions covering the ranges of ambient temperature, altitude, voltage variation, frequency variation, mounting, and ventilation within which motors are intended to operate unless specific exceptions to these conditions are agreed to. In general, satisfactory operation can be expected within the specified ranges, but for best efficiency the motor should be operated near its nameplate values of load, voltage, and frequency. It would be almost impossible to state a generalized rule for the effect of voltage and frequency variations on efficiency that would be applicable to the wide range of motor designs which are available. If the voltage is raised too far above its nominal value, magnetic saturation may occur, magnetizing current increases, and iron losses increase. If the voltage is reduced below nominal, slip increases and the secondary $I^2r$ losses go up. Frequency changes have similar effects but in the opposite sense. Lowering the frequency without a corresponding change in voltage causes magnetic saturation and increasing frequency without a voltage change causes slip to increase. The best practice is to maintain voltage and frequency at the motor's rated values and not be satisfied with staying within the specified ±10% and ±5% variations for voltage and frequency, respectively.

Most motors are rated for 40°C ambient operation but better efficiency will be realized if the ambient is lower. Again, a general rule is difficult to formulate because grease and oil stiffen at very low temperatures causing an increase in friction losses. The relationship of grease viscosity to temperature is quite different for the different types of greases used in motors of various manufacturers; however, all greases in common usage will operate satisfactorily in the range of 0–40°C.

Proper attention to providing adequate ventilation is another means of ensuring that the motor operates within its design temperature range and, therefore, will provide its expected efficiency.

In summary, application conditions have a strong influence on motor losses. Even an energy-efficient motor, if improperly applied, may not produce the expected savings, and correcting improper service conditions may result in as much or more savings than replacing existing motors.

## 3.10. Keeping Motors Running

Electric motors operate under a wide variety of adverse conditions. In general, they require little maintenance. Most motors are quite robust and usually will endure considerable mistreatment, while still continuing to perform their assigned task. However, a little attention to good application practices at the time of the initial motor installation and to periodic maintenance will aid in realizing the full benefits from installing an energy-efficient motor. Generally, the initial investment in an energy-efficient motor exceeds the investment that would have been made for a standard motor of equivalent rating, so there is added incentive to keep the motor running. Here are a few tips that will help.

### Service Factor and Its Proper Use

Many motors are provided with a service factor greater than 1.0. This service factor indicates that the motor has a safety factor which will enable it to carry overloads up to the service-factor rating without failure. A common service factor for the medium size range of ac motors is 1.15. This is sometimes misinterpreted to mean that the motor can carry 15% greater than its nameplate horsepower rating without any deterioration in life. Such an interpretation simply is not true. The NEMA Standards[1] provide that a general-purpose ac motor with a service factor of 1.15 must operate continuously at the service-factor load with a 90°C temperature rise measured by the resistance method. The equivalent temperature rise for a 1.0 service-factor machine is 80°C. This 10°C differential, if utilized on a continuous basis, will reduce the insulation life of a motor.

Service factors greater than 1.0 should be utilized for their intended purpose, that is, to accommodate unavoidable overload conditions of short duration. Dependence on the service factor to accommodate a continuous overload condition may result in a disappointing service life from the motor.

## Alignment

Misalignment causes increased bearing loading and frictional losses. Alignment by eyeball and feel, rather than gauges, can result in loss of efficiency and shortened motor life. Assuming alignment is correct at the time of installation, it should be checked periodically, as any of the following can cause the alignment to change:

- Vibration
- Temperature changes
- Stress from connecting pipelines
- Wear

## Belt Tension

Belted applications need periodic attention. Loose belts cause slippage, and overtightened belts overload the motor's bearings. Both situations waste energy and the overtightened belt can, in addition, cause bearing failures.

## Lubrication

Periodic lubrication is required to prolong bearing life of motors with grease-lubricated bearings. The practice which is sometimes followed of pumping as much grease as possible into the bearing, is not consistent with obtaining the best efficiency. Overgreasing causes the friction losses to increase. This is particularly true for 3600-rpm motors. In some motor designs, overlubrication can force grease out of the bearing housings and cause it to be deposited onto the motor windings. This can eventually cause insulation failure. The motor manufacturer's greasing instructions should be followed to avoid either premature bearing failure for lack of grease, or wasted energy expended in churning excess grease.

## Ventilation

An enclosed motor must dissipate the heat generated by its internal losses through its frame and brackets to the outside air either by convection or radiation. Most enclosed motors have finned frames to increase the available dissipating area. In contaminated environments, blankets of dust, oil, mist, or other insulating substances, can cover the frame and fins. This provides an added thermal barrier through which the motor's internal heat must pass to reach the outside air and, in some cases, the particle buildup also reduces the volume of air passing across the frame's cooling fins. It is not uncommon to ignore this condition and motors are frequently robust enough to provide many years of service even under such adverse conditions.

It is a misconception to believe that a motor's operating temperature can be judged to be satisfactory or not by feeling the frame with your hand. The average person cannot stand more than about 125 °F without discomfort and the frame temperatures of many Design B motors operating within the allowable temperature rise will considerably exceed this temperature.

Beyond simply extending the motor's life, there is now added incentive to periodically clean motors. The increased thermal resistance resulting from accumulations on the motor frame causes the temperature rise of the machine to increase. This causes an increase in both primary and secondary $I^2r$ losses. In the 50-hp example given in Table 2.1, an increase of 20 °C in the average internal temperature of the motor would reduce the efficiency by 0.2%, a needless waste of energy.

Weatherproof protected motors, which are commonly used in the large ac sizes for motors installed outside, draw their cooling air through filters. If these filters become clogged, the cooling air supply is reduced and the motor's internal temperature goes up. Dripproof motors may also have their air passages clogged over time in particularly bad environments, causing reduced airflow and a higher temperature rise in the motor. An ample supply of cool air is needed to keep the motor operating at its rated efficiency.

## Moisture

Moisture causes insulating materials to lose their insulating properties. Mapes[10] cites the example of a motor in a petrochemical plant which evidenced a drop in insulation resistance from 500 to 1 MΩ after 1 week of storage.

In damp, humid locations, motor windings are subjected to condensation formation each time the motor is stopped, and the temperature of the windings drops below the dew point. Even totally enclosed and explosionproof motors breathe. As the motor cools down, air with moisture is drawn into the motor. Over a period of time, this repeated condensation buildup will degrade the insulation and may eventually cause a breakdown.

The same problem can occur, of course, when motors stand idle in storage in a high-humidity environment.

A solution to this problem is to maintain the temperature of the motor windings above the dew point, thus preventing condensation. There are two common methods of doing this:

1. Space heaters. Resistance-type heaters are mounted inside the motor. The power is controlled so that the heater comes on when the motor is off.
2. Low-voltage heating. Single-phase voltage is applied directly to the motor windings when the motor is idle. The single-phase power can be obtained from a step-down transformer or from a solid-state controller which reduces the voltage applied to the motor windings by phase-controlled thyristors.

None of these precautions are new. Most are familiar to people involved with motor installation and maintenance. However, the increased cost of energy and the operating dollars which are wasted when motors, through improper maintenance, do not continue to operate at their designed efficiency provide a new incentive for proper application and maintenance practices.

## References

1. NEMA Standards Publication No. MG1, "Motors and Generators." Washington, D.C.: National Electrical Manufacturers Association, 1993.
2. Veinott, C.G. *Theory and Design of Small Induction Motors.* New York: McGraw–Hill Book Co., 1959.
3. Rawcliffe, G.H., Burbridge, R.F., and Fong, W. Induction-motor speed changing by pole-amplitude modulation, *Proceedings IEE,* Vol. 105, Part A, August, 1958.
4. Rawcliffe, G.H., and Fong, W. Speed-changing induction motors, *Proceedings IEE,* Vol. 107, Part A, December, 1960.
5. Rawcliffe, G.H., and Fong, W. Close-ratio two speed single-winding induction motors, *Proceedings IEE,* Vol. 110, May, 1963.
6. Alger, P.L. *The Nature of Polyphase Induction Machines.* New York: John Wiley & Sons, Inc., 1951.
7. Clarke, E. *Circuit Analysis of AC Power Systems, Volume I: Symmetrical and Related Components.* New York: John Wiley & Sons, Inc., 1943.
8. Wagner, C.F., and Evans, R.D. *Symmetrical Components.* New York: McGraw–Hill Book Co., 1933.
9. Nailen, R.L. The hidden costs of electric motor cycling, *Plant Engineering,* March 22, 1979.
10. Mapes, W.H. How to keep motors running, *Chemical Engineering,* July 21, 1975.

# 4

# Efficiency Labeling and Verification

## 4.1. Variations in Motor Efficiency

A plant engineer installs a 100-hp motor on a new processing line being started up in his plant. He determines from the motor's nameplate that its efficiency is 93.0%. Can he conclude that if he operates the motor with rated voltage and frequency applied and at 100-hp load the motor will use precisely $(100 \times 0.746)/0.93 = 80.215$ kW? The answer, of course, is no. Then what can this plant engineer expect to experience in the way of power consumption? What are the tolerances and factors that must be considered? This chapter discusses the statistical nature of motor efficiency, testing methods for verifying efficiency, and the tolerance limits on efficiency that the motor user can expect.

Motors, like refrigerators, automobiles, and many other products, are an assembly of many different parts and materials, each of which individually must meet a specification which includes a tolerance on its physical dimensions and/or physical characteristics. Therefore, all motors of a particular horsepower rating and model number made by the same motor manufacturer do not have identical efficiencies. However, NEMA[1] has standardized an efficiency labeling procedure so that the motor user has a clearly defined range of efficiency variations that he can expect when he installs a NEMA efficiency labeled motor. Another option that is sometimes utilized is to purchase a motor with a guaranteed minimum efficiency and to include as part of the purchase agreement that the motor will be tested with either a certified copy of the test report provided or the test witnessed by a representative of the motor buyer. Because of the expense, this option of individually guaranteeing, testing, and reporting is usually reserved for larger horsepower motors, approximately 1000 hp and up.

Now, let us consider some of the variables which contribute to efficiency variations. One important influence is the measurement accuracy limitations of motor testing. To obtain an overview of the nature of this problem, consider the following typical data used to determine the efficiency of a 100-hp motor. Only the input kilowatt measurement will be considered in this example as this is sufficient to illustrate the nature of the measurement problem. However, additional data are required to determine a motor's efficiency and measurement accuracy limitations exist for the other measured variables as well.

Table 4.1 illustrates how a 1/2% of full-scale error in the power input measurement causes the true value of efficiency, namely 92.5%, to be reported as 93.1%. This 1/2% measurement error, which is not unusual for analog-type wattmeters, caused an error in the observed motor losses of 8.3%. This arises from the fact that efficiency, calculated by dividing output by input, requires the measurement of two quantities, both of which are large with respect to the motor losses. Small errors in either of these measurements produce a disproportionately large error in the difference between the two quantities, which is the motor loss. This subject has undergone extensive investigation[2,3] and means for greatly improving testing accuracy are discussed later in this chapter.

**Table 4.1. Wattmeter Error Effect on Motor Efficiency**

| | Reading | Wattmeter error[a] | True value |
|---|---|---|---|
| Input power, kW; polyphase wattmeter × instrument transformer ratios | 80.129 | 0.500 | 80.629 |
| Output power, kW | 74.6 | — | 74.6 |
| Losses = Input − Output, kW | 5.529 | | 6.029 |
| $\text{Efficiency (\%)} = \dfrac{\text{Output power} \times 100}{\text{Input power}}$ | 93.1 | — | 92.5 |

[a] Calibration error is 1/2% of full-scale reading, and the full-scale reading for the wattmeter and instrument transformer combination is 100 kW.

The second contributing factor to efficiency variation is the differences that occur in motor characteristics even in motors built to the same design. These differences are caused by changes in the raw materials used in the motor, such as copper, steel, and aluminum, and also by manufacturing process variations. Although all may be within specified tolerances, there will still be some differences in motor losses among motors of a duplicate design. It is common practice to assign a single value of efficiency to all motors of one design. The range of efficiencies that can actually be expected from motors out of this design population is the subject of the next section.

## 4.2. Efficiency Labeling

Mindful of the tolerances that apply to any statement of a motor's efficiency, NEMA[1] created an efficiency labeling standard which recognizes the realities of motor-to-motor variations and testing accuracy. This standard gives the motor user well-defined efficiency information which he needs to select and apply motors. Consider a popular model of motor which is manufactured in large quantities. If the number of motors having a certain value of efficiency were plotted against efficiency values and a curve drawn through these points, the curve would approximate the familiar bell-shaped curve referred to in statistics as a Gaussian distribution. Figure 4.1 displays such a curve.

The mean value of efficiency for the population of motors is given by

$$\bar{\eta} = \frac{1}{n} \sum_{k=1}^{n} \eta_k \tag{4.1}$$

where $\bar{\eta}$ = mean value of efficiency, $\eta_k$ = efficiency of the $k$th motor, and $n$ = total number of motors in the population.

Another important parameter used in statistics to describe the distribution of the population about the mean is the standard deviation, $\sigma$:

$$\sigma = \sqrt{\frac{1}{n} \sum_{k=1}^{n} (\eta_k - \bar{\eta})^2} \tag{4.2}$$

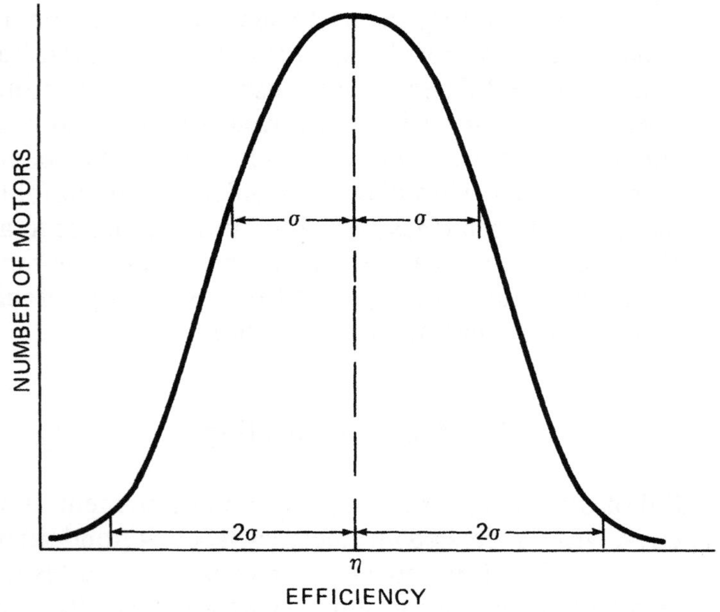


**Figure 4.1.** Gaussian distribution of motor efficiencies.

The mean, $\sigma$, and $2\sigma$ values are shown in Fig. 4.1. The band encompassed by $\bar{\eta} \pm \sigma$ contains 68.3% of the population and the band encompassed by $\bar{\eta} \pm 2\sigma$ contains 95.4% of the population.

With these concepts in mind, the logic of defining motor efficiency in terms of an efficiency band rather than a discrete number is evident. NEMA has published a set of tables defining nominal and minimum efficiencies for both standard and energy-efficient motors. One of these tables is shown here as Table 4.2. For each efficiency band three quantities are defined:

- Nominal efficiency
- Minimum efficiency based on 20% loss difference
- Minimum efficiency based on 10% loss difference

The nominal efficiency corresponds to the mean value, $\bar{\eta}$, for the band and is the value to be used in calculating energy usage. While any given motor may not have an efficiency exactly equal to the nominal value, one can expect to find as many motors with efficiencies above as below the nominal value and thus the average energy usage will be predicted correctly using the nominal efficiency value. The minimum

**Table 4.2. Table of Nominal and Minimum Efficiencies**[a]

| Column A Nominal Efficiency | Column B (NEMA Standard) Minimum Efficiency Based on 20% Loss Difference | Column C (Suggested Standard for Future Design) Minimum Efficiency Based on 10% Loss Difference |
|---|---|---|
| 99.0 | 98.8 | 98.9 |
| 98.9 | 98.7 | 98.8 |
| 98.8 | 98.6 | 98.7 |
| 98.7 | 98.5 | 98.6 |
| 98.6 | 98.4 | 98.5 |
| 98.5 | 98.2 | 98.4 |
| 98.4 | 98.0 | 98.2 |
| 98.2 | 97.8 | 98.0 |
| 98.0 | 97.6 | 97.8 |
| 97.8 | 97.4 | 97.6 |
| 97.6 | 97.1 | 97.4 |
| 97.4 | 96.8 | 97.1 |
| 97.1 | 96.5 | 96.8 |
| 96.8 | 96.2 | 96.5 |
| 96.5 | 95.8 | 96.2 |
| 96.2 | 95.4 | 95.8 |
| 95.8 | 95.0 | 95.4 |
| 95.4 | 94.5 | 95.0 |
| 95.0 | 94.1 | 94.5 |
| 94.5 | 93.6 | 94.1 |
| 94.1 | 93.0 | 93.6 |
| 93.6 | 92.4 | 93.0 |
| 93.0 | 91.7 | 92.4 |
| 92.4 | 91.0 | 91.7 |
| 91.7 | 90.2 | 91.0 |
| 91.0 | 89.5 | 90.2 |
| 90.2 | 88.5 | 89.5 |
| 89.5 | 87.5 | 88.5 |
| 88.5 | 86.5 | 87.5 |
| 87.5 | 85.5 | 86.5 |
| 86.5 | 84.0 | 85.5 |
| 85.5 | 82.5 | 84.0 |
| 84.0 | 81.5 | 82.5 |
| 82.5 | 80.0 | 81.5 |
| 81.5 | 78.5 | 80.0 |
| 80.0 | 77.0 | 78.5 |
| 78.5 | 75.5 | 77.0 |
| 77.0 | 74.0 | 75.5 |
| 75.5 | 72.0 | 74.0 |
| 74.0 | 70.0 | 72.0 |
| 72.0 | 68.0 | 70.0 |
| 70.0 | 66.0 | 68.0 |
| 68.0 | 64.0 | 66.0 |
| 66.0 | 62.0 | 64.0 |
| 64.0 | 59.5 | 62.0 |
| 62.0 | 57.5 | 59.5 |
| 59.5 | 55.0 | 57.5 |
| 57.5 | 52.5 | 55.0 |
| 55.0 | 50.5 | 52.5 |
| 52.5 | 48.0 | 50.5 |
| 50.5 | 46.0 | 48.0 |

[a] This is Table 12-8 of NEMA Standards Publication MG 1-1993. It is reproduced by permission of the National Electrical Manufacturers Association which holds the copyright.

efficiency represents the lowest value any motor which has been properly labeled for a particular band can have. For example, if a group of 10-hp motors are labeled with a nominal efficiency of 88.5%, all members of the group must have efficiencies equal to or greater than 86.5% based on Column B in Table 4.2. Therefore, the motor user has a specific basis for acceptance or rejection of any individual motor. The NEMA labeling standard is defined as being applicable to motors in the 1–125 hp range so it encompasses the segment of the motor population which accounts for the major portion of motor energy usage as discussed in Chapter 1. Keinz and Houlton[4] have described the testing and analysis that preceded the establishment of the efficiency bands and have detailed the merits of the efficiency band approach for specifying motor efficiencies.

## 4.3. Testing of Induction Motors

IEEE Standard No. 112 is the U.S. standard which defines testing procedures for polyphase induction motors. The standard provides several alternate methods for measuring efficiency.*

### Method A: Input–Output

A mechanical brake or dynamometer is used to load the motor and the reaction is measured to determine the motor's torque. Efficiency is calculated as output/input.

The resulting calculated efficiency is subject to the measurement errors described earlier in this chapter.

### Method B: Input–Output with Loss Segregation

A dynamometer is used to load the motor. The motor torque is measured by load cell or scale. Most dynamometers have electrical controls for setting the load torque and holding it steady while readings are being taken. This method is the method recommended by NEMA[1] for efficiency measurements in the 1- to 125-hp sizes.

* Please note that Methods A through C, E, and F are outlined in the following sections. The omission of Method D is deliberate. IEEE 112 does not define a Method D.

The accuracy of this test method for calculating motor efficiency is considerably improved over Method A. The reason is that the losses are evaluated individually and then summed to arrive at the total motor loss. This procedure mitigates the difficulty inherent in determining a small number, losses, as the difference between two large numbers, input and output power. Later in this chapter a more detailed discussion of the loss segregation technique is presented.

### Method C: Duplicate Machines

Two identical machines are coupled together and supplied from two separate power sources. One of the two sources must be an adjustable-frequency source. Controlling the adjustable-frequency source to operate at frequencies above and below the fixed-frequency source causes each machine to operate as a motor and then as an induction generator returning power back to the line. Electrical power into and out of the two machines is measured and the difference is the combined loss of the two machines. This method has the advantage that the net power consumed is equal to the losses of the two machines, and the majority of the power entering the system is returned to the line through the machine which is acting as an induction generator. Two identical machines must be available for test by this method and the adjustable-frequency supply must have a rating equal to or greater than the machines being tested.

IEEE 112[5] describes a loss segregation technique analogous to the one used for Method B to arrive at the corrected total motor losses and thus the motor efficiency.

### Method E: Input Measurement

This method requires that the motor being tested must be loaded, but measuring the motor output is not required. Readings of the voltage, current, electrical power input, speed, temperature, and winding resistance are taken. From these data, four of the motor's five losses can be calculated but the fifth, namely the stray-load loss, requires separate tests. These separate tests, called reverse-rotation and rotor-removed tests, are needed to directly measure the stray-load loss. The five losses are then combined along with the input power measurement to determine the motor's efficiency.

An alternative to conducting the stray-load loss tests is permitted[5] in which the stray-load loss is assumed to be a specified percentage of the output.

## Method F: Equivalent Circuit

This method does not require any loading device although an optional loaded test is available under Method F. In one of the options provided, the motor's equivalent circuit parameters are measured by taking tests at no load and locked rotor. Figure 4.2 is the equivalent circuit used for Method F. The stator resistance, $r_1$, is measured by a resistance bridge or equivalent means. The leakage reactances, $x_1$ and $x_2$, are determined from an impedance test taken at reduced frequency. The rotor is locked and the frequency is reduced to a maximum of 25% of rated frequency. Likewise, $r_2$ is determined from the reduced frequency locked-rotor test. This reduced frequency test is required to avoid distorting the values of $r_2$ and $x_2$ by the skin effect that occurs in rotor bars of most medium and large ac motors at 60 Hz. Of, course, the reactances, $x_1$ and $x_2$, must be corrected to rated frequency for use in the equivalent circuit calculations.

Alternative test procedures for measuring $r_2$ and $x_2$ are provided in IEEE 112.[5] A load point test is one of the alternatives. In this method the motor is operated at a slip near the expected rated load slip by reducing the motor voltage at no load or loading the motor plus reducing the motor voltage as required. The values of $x_m$ and $r_{fe}$ are determined from data taken during the no-load and impedance tests.

Details of the various test methods for determining the induction motor equivalent circuit parameters are given in IEEE 112.[5]

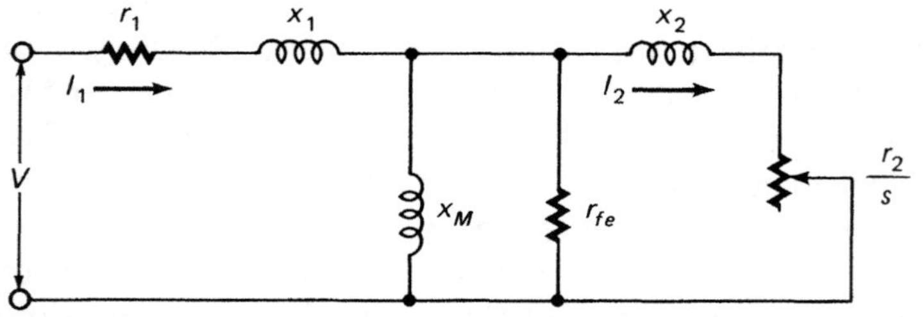


**Figure 4.2.** Induction motor equivalent circuit for Method F calculation.

Once the equivalent circuit constants have been determined, the efficiency is calculated for any load by solving the equivalent circuit. This is an iterative procedure in which one assumes a value of slip, $s$, calculates the output, and then tries a new value of slip until the output is sufficiently close to the desired value. Stray-load loss as well as friction and windage must be added to the equivalent circuit losses. The procedure for determining stray-load loss given under Method E is also applicable to Method F.

## Additional Test Methods: Equivalent Circuit Calibrated with One Load Point

When a slip current point is available under load, this information can be used to improve the estimate of the equivalent circuit parameters, primarily $r_2$. The equivalent circuit technique for calculating efficiency is then used with the improved estimate of $r_2$.

Each of these various test methods has its own particular realm of applicability. It would be desirable to have a single, best test method applicable to all motors, but, unfortunately, this is not practical. There are some types of motors for which loading by a dynamometer is not possible. A large vertical machine which cannot be operated in a horizontal position is an example. Also, loading devices such as dynamometers are expensive and are not always available in the sizes needed to test large-horsepower motors. The mechanical brake, permitted in Method A, is a practical way to load a fractional-horsepower motor but is completely impractical for testing a 500-hp machine. Hence, a variety of test methods are needed and are available in IEEE 112.[5]

Recognizing the concentration of energy usage in the 1–125 hp range, NEMA has selected Method B, dynamometer, as the recommended method for efficiency testing machines in this horsepower category. Dynamometers of these sizes are readily available, and, unless some motor construction feature inhibits its being coupled to a dynamometer, Method B is recommended for use.

The problem of accurately determining the motor losses which are the difference between two large numbers, input and output power, has been discussed earlier in this chapter. Table 4.1 demonstrates how a relatively small error in reading the input power has an exaggerated effect on the observed value of motor losses and thus the efficiency. This suggests that testing accuracy can be improved if the test is constructed so that the losses can be individually determined. This "segregation of losses" technique, described below, provides the most accurate method available for determining motor efficiency. In the description which follows, it is assumed that a no-load saturation test and a dynamometer test, Method B, have been conducted. The procedure described uses these data to segregate the losses, the data are used to conduct some data smoothing to minimize random errors, and then the motor efficiency is calculated. The detailed steps in the data reduction process are given in IEEE 112.[5] This discussion is an overview of the procedure to provide an understanding of the basic steps one goes through in reducing the raw test data to the final value of motor efficiency. Some of the refinements, such as temperature corrections, dynamometer corrections, and instrument calibrations, are omitted for the sake of clarity. The complete algorithm for performing the data reduction is extensive and, therefore, is usually performed by a computer.

## Iron Loss and Friction and Windage from No-Load Saturation Test

The no-load saturation test is conducted by measuring voltage, current, power, and winding temperature for several values of applied voltage. A typical curve of power and current plotted versus voltage is displayed in Fig. 4.3. Friction and windage is determined by using the lower values of voltage from the no-load saturation test and plotting watts versus (volts)$^2$ as shown in Fig. 4.4. Extrapolating this line to zero volts gives the value of friction and windage loss.

Now iron loss, $W_{Fe}$, can be calculated by

$$W_{Fe} = W_0 - KI_0^2 r_1 - W_{FW} \tag{4.3}$$

where $W_0$ = power (see Fig. 4.3), $I_0$ = current (see Fig. 4.3), $r_1$ = stator winding resistance, $K$ = conversion factor to obtain $I_0^2 r_1$ for all three phases, 1.5 if $r_1$ is line-to-line resistance, and $W_{FW}$ = friction and windage loss (see Fig. 4.4).

The iron loss is really more a function of the air-gap induced voltage in the machine than it is of the terminal voltage. Refined test analysis methods take this into account by developing a curve of iron

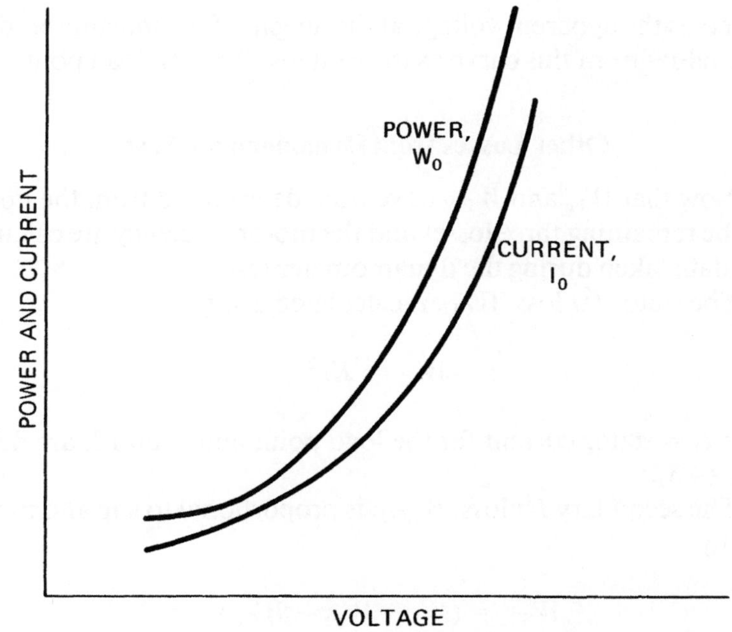


**Figure 4.3.** No load saturation curve for induction motor.

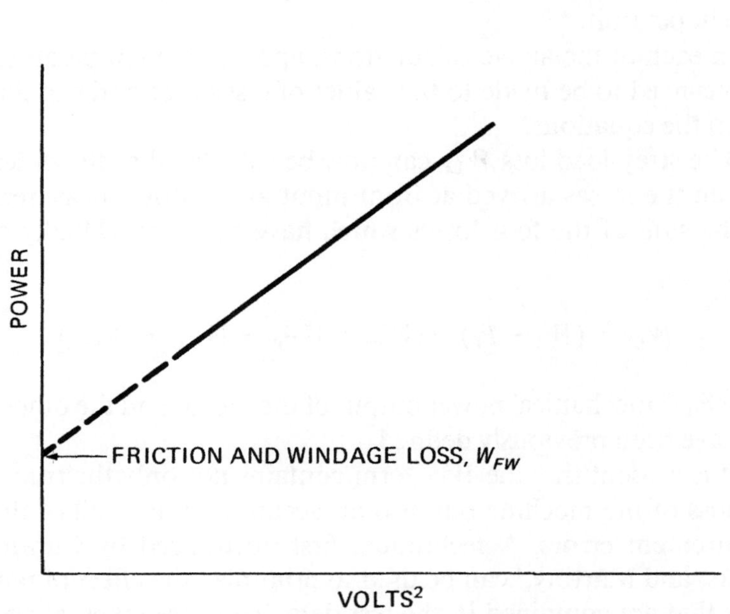


**Figure 4.4.** Determination of friction and windage loss from no-load saturation test.

loss versus the apparent voltage at the air gap of the machine and then using values from this curve as the iron loss for each load point.

## Other Losses from Dynamometer Test

Now that $W_{Fe}$ and $W_{FW}$ have been determined from the no-load test, the remaining three losses and the motor efficiency are calculated from data taken during the dynamometer test.

The stator $I^2r$ loss, $W_{CU}$, is calculated using

$$W_{CU} = KI_1^2 r_1 \tag{4.4}$$

where $I_1$ = stator current for the load point and $r_1$ and $K$ are defined in Eq. (4.3).

The secondary $I^2r$ loss, $W_{SEC}$, is proportional to slip and is calculated by

$$W_{SEC} = (W_1 - W_{Fe} - W_{CU}) \times s \tag{4.5}$$

where $W_1$ = total watt input for the load point being considered and $s$ = slip in per unit.

In each of the above calculations, appropriate temperature corrections need to be made to the values of resistance and slip that are used in the equations.

The stray-load loss $W_{LL}$ can now be calculated as the difference between the losses arrived at from input and output measurements and the sum of the four losses which have been individually calculated:

$$W_{LL} = (W_1 - P_0) - (W_{Fe} + W_{FW} + W_{CU} + W_{SEC}) \tag{4.6}$$

where $P_0$ = mechanical power output of the motor and the other symbols have been previously defined.

It is evident that the $W_{LL}$ term contains not only the true stray-load loss of the machine but also an accumulation of all of the test measurement errors. A technique, first introduced by Cummings, Bowers, and Martiny,[3] can be used to minimize the effect of random errors that are contained in the test data. Tests are taken at six load points evenly spaced between 1/4 and 6/4 rated load. For each load point the losses are segregated as explained above and a value of stray-load loss is calculated using Eq. (4.6). The stray-load loss points are then plotted versus torque squared as shown in Fig. 4.5. Theoretically, these points should plot as a straight line since stray-load loss is considered to vary as motor torque squared. A straight line is then mathematically plotted through the test points using a linear regression[6] curve fit. The equation for this straight line is of the general form

$$W_{LL} = AT^2 + B \tag{4.7}$$

Now, the stray-load loss should be zero when the motor torque is zero, so the straight line is shifted to have a zero intercept on the "$y$" axis. This is shown graphically in Fig. 4.5 and mathematically the equation for the corrected value of stray-load loss is

$$W_{LLC} = AT^2 \tag{4.8}$$

The stray-load loss value from (4.8) is then used along with the other four losses to calculate the motor's efficiency using

$$\text{Efficiency} = \frac{W_1 - (W_{LLC} + W_{CU} + W_{Fe} + W_{FW} + W_{SEC})}{W_1} \tag{4.9}$$

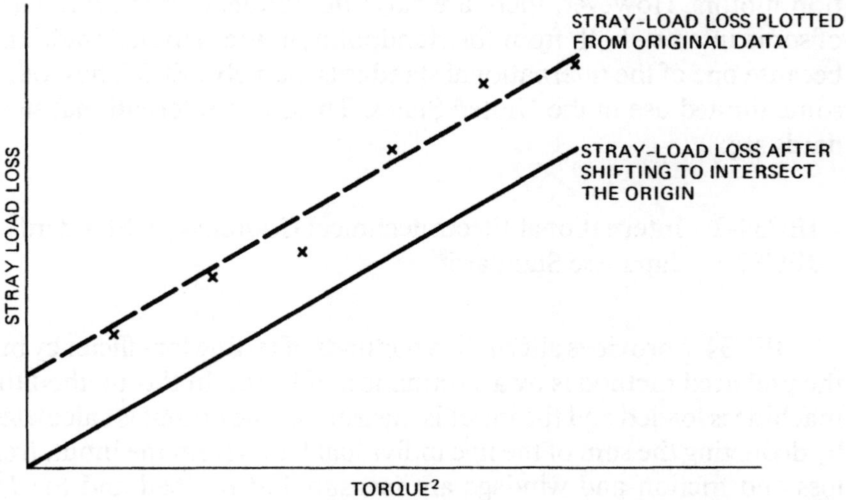


**Figure 4.5.** Linear regression curve fit of stray-load loss test data.

This process of "smoothing" the stray-load losses has the effect of minimizing the random errors that are present in the test data. It will not, of course, take a bad test, carelessly run with uncalibrated instruments, and make a good test out of it. There is a protective feature built into the test procedure which specifies that if the correlation factor,[6] $r$, which tests the "goodness of fit" of the straight line to the test data, is less than 0.9, 1.0 being a perfect fit, one point may be deleted. Then the values for $A$ and $B$ in (4.7) along with a new correlation factor are recalculated. If $r < 0.9$ with one point deleted, the test data are declared unacceptable and the test must be repeated after the cause of the test errors is corrected. IEEE 112[5] provides additional guidance on reducing the measurement errors and improving the test correlation factor, $r$.

As noted earlier in this section, the reduction of test data to a final value of efficiency is considerably more complex than one might at first envision but the improvements in testing accuracy which result are well worth the effort.

## 4.4. International Testing Standards for Motors

IEEE 112[5] is the American National Standard for testing induction motors. However, there are two international testing standards of some interest, both from the standpoint of international trade and because one of the international standards, namely JEC37, has found some limited use in the United States. These two international standards are:

| | |
|---|---|
| IEC34-2 | International Electrotechnical Commission Standard[7] |
| JEC37 | Japanese Standard[8] |

IEC34-2 provides alternative methods of testing for efficiency but the preferred method is by a summation of losses. In this method the machine is loaded and the input is measured. The output is calculated by deducting the sum of the five individual losses from the input. Iron loss and friction and windage are measured at no load and the $I^2r$ losses are calculated from the test data. Stray-load loss is an assumed value, 0.5% of the rated input for motors and 0.5% of the rated output for induction generators.

JEC37 utilizes a circle diagram solution as the principal method for determining efficiency. Tests taken at no load and locked rotor provide the data for constructing the circle diagram. The JEC37 method does not include any explicit value for stray-load loss. The stray load loss is not measured as in the IEEE test code nor assumed as in the IEC test code.

Jordan and Gattozzi[2] reported the results of four tests for motor sizes ranging from 15 to 1500 hp in which the IEC, JEC, and IEEE test methods were compared. The results of these tests are summarized in Table 4.3. In all but one test, the IEEE Method B test results yielded the lowest efficiency. Cummings *et al.*[3] reported 15 tests covering a range from 350 to 1250 hp. Comparing Cummings's data in the col-

**Table 4.3. Comparison of Different Test Methods for Efficiency Determination**

| | Test method | | |
|---|---|---|---|
| | IEEE 112 Method B | JEC37 | IEC34-2 |
| | Test load—15 hp | | |
| Current, A | 19.2 | 18.7 | 19.2 |
| Efficiency, % | 87.4 | 90.1 | 89.2 |
| Power factor, % | 84.4 | 83.5 | 84.5 |
| | Test load—70 hp | | |
| Current, A | 86.1 | 81.6 | 86.1 |
| Efficiency, % | 90.0 | 93.1 | 92.7 |
| Power factor, % | 86.2 | 86.3 | 86.2 |
| | Test load—800 hp | | |
| Current, A | 99.5 | 97.8 | 101.3 |
| Efficiency, % | 95.9 | 95.9 | 95.6 |
| Power factor, % | 89.6 | 91.9 | 89.6 |
| | Test load—1500 hp | | |
| Current, A | 188.5 | 185.7 | 188.5 |
| Efficiency, % | 95.9 | 96.8 | 96.0 |
| Power factor, % | 87.5 | 89.8 | 87.5 |

umn marked E/F in the Cummings *et al.* paper[3] (referring to the IEEE test method) with tests on the same motors by the JEC and IEC test methods, the same conclusion is reached. The IEC and JEC test methods generally yield efficiency values higher than IEEE 112. There are several differences in testing techniques which contribute to the discrepancy but the major factor is the different methods of handling stray-load loss. JEC37 does not include stray-load loss as a component of the total loss and the IEC34-2 assumes a fixed percentage for stray-load loss. IEEE 112, Method B measures stray-load loss which, in the author's view, is the best method and yields a value of motor efficiency more representative of that which will be experienced under actual application conditions.

## References

1. NEMA Standards Publication No. MG1, "Motors and Generators." Washington, D.C.: National Electrical Manufacturers Association, 1993.
2. Jordan, H.E., and Gattozzi, A. Efficiency testing of induction machines, *IEEE Industry Applications Society Conference Record,* 1979.
3. Cummings, P.G., Bowers, W.D., and Martiny, W.J. Induction motor efficiency test methods, *IEEE Industry Applications Society Conference Record,* 1979.
4. Keinz, J.R., and Houlton, R.L. NEMA nominal efficiency, what is it and why? *IEEE Petroleum and Chemical Industry Conference Record,* September, 1980.
5. IEEE 112, "IEEE Standard Test Procedure for Polyphase Induction Motors and Generators." New York: The Institute of Electrical and Electronics Engineers, 1991.
6. Ezekile, M., and Fox, K.A. *Methods of Correlation and Regression Analysis.* New York: John Wiley and Sons, Inc., 1959.
7. "Rotating Electrical Machines, Part 2, Methods for Determining Losses and Efficiency of Rotating Electrical Machinery from Test," Publication 34-2. Geneva, Switzerland: International Electrotechnical Commission, 1972.
8. "Standard of the Japanese Electrotechnical Committee, Induction Machine," JEC37. Denki Shoin, Tokyo, Japan.

# 5

# Power Factor

## 5.1. Power Factor and Its Effect on Energy Consumption

In an ac motor two components of energy are present. One component is the real power which is converted from electrical into mechanical energy and into heat in the form of motor losses. The second, reactive energy, is stored in the inductive elements of the machine. If a motor, or any ac circuit for that matter, has no capacitive or inductive elements, then all of the energy is expended as real power, but if either capacitance, inductance, or both are present, then some reactive energy will also be present. Electric motors contain inductive elements and, therefore, reactive energy is a consideration in selecting and applying motors in an energy-efficient manner.

Figure 5.1 shows voltage and current waveforms for a simple ac electrical circuit consisting of a resistance R and inductance L. The current is said to lag behind the voltage for the case illustrated in this figure because the positive (or negative) peak in the current waveform occurs at some point in time later than the peak of the voltage sine wave. The real power can be calculated by multiplying the voltage and the current, $v \times i$. At any particular instant of time, such as point A in Fig. 5.1, the instantaneous power, $p_i$, is given by

$$p_i = v_i \times i_i \tag{5.1}$$

The average power is obtained by averaging the instantaneous values of the $p_i$'s over one complete cycle, for example, from 0 to 360°. Intu-

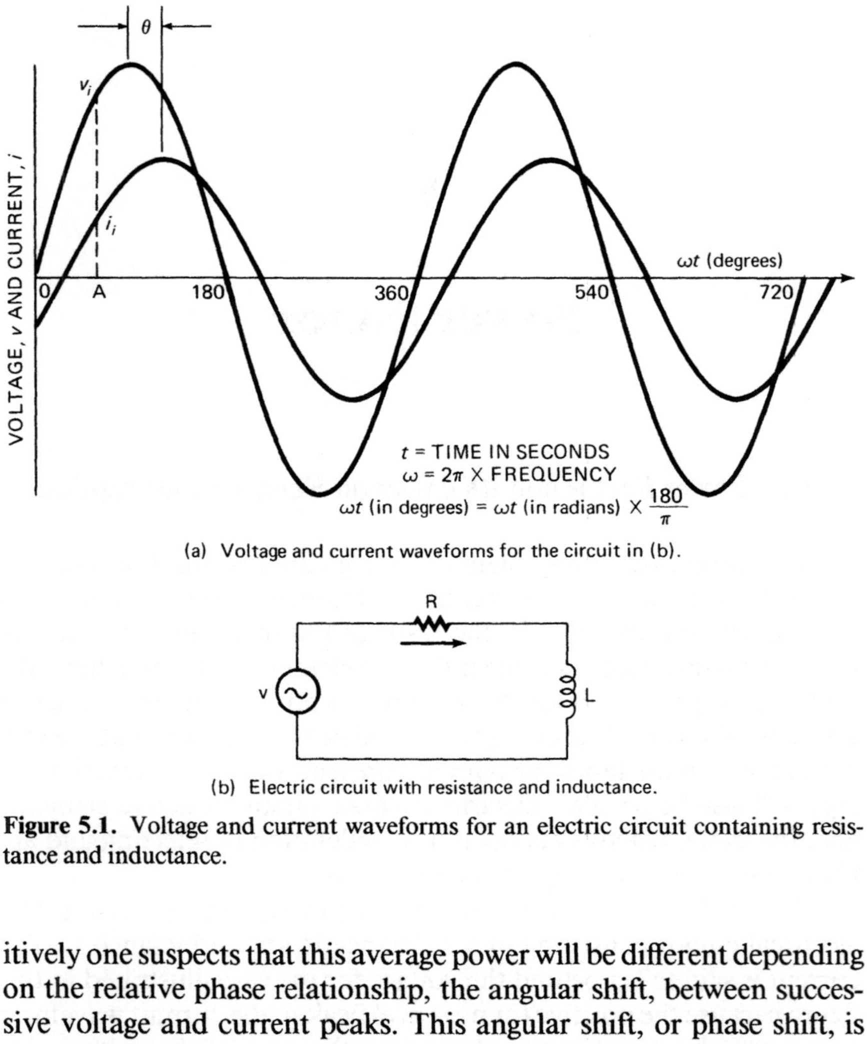


**Figure 5.1.** Voltage and current waveforms for an electric circuit containing resistance and inductance.

itively one suspects that this average power will be different depending on the relative phase relationship, the angular shift, between successive voltage and current peaks. This angular shift, or phase shift, is shown in Fig. 5.1 as the angle $\theta$. The two extreme cases in Fig. 5.2(a) and (b) make this clear. In Fig. 5.2(a) the $v_i \times i_i$ products will all be positive in the segment between u and w and they will all be negative between w and x. The average power over this half cycle will be zero. Similarly the average power over the second half cycle x, y, and z will also be zero. In Fig. 5.2(b) the $v_i \times i_i$ products are all positive and the real power is a maximum for this phase relationship between the voltage and current. Tang's[7] and many other textbooks on ac circuits derive the relationships between voltage, current, and power so the

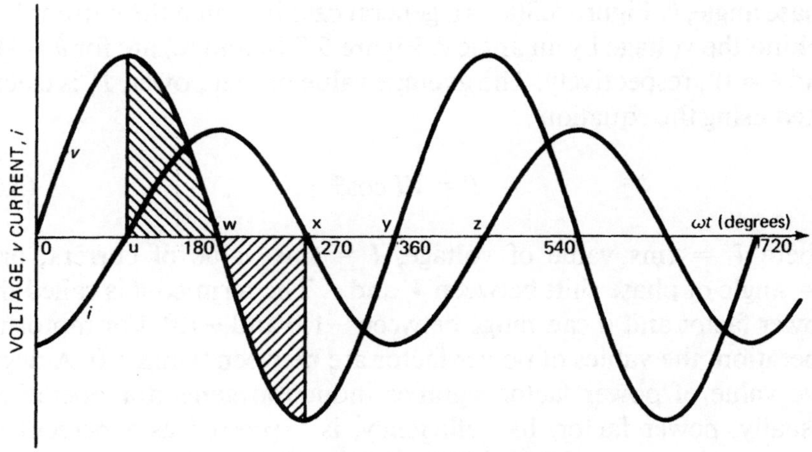


(a) Voltage and current 90° out of phase.

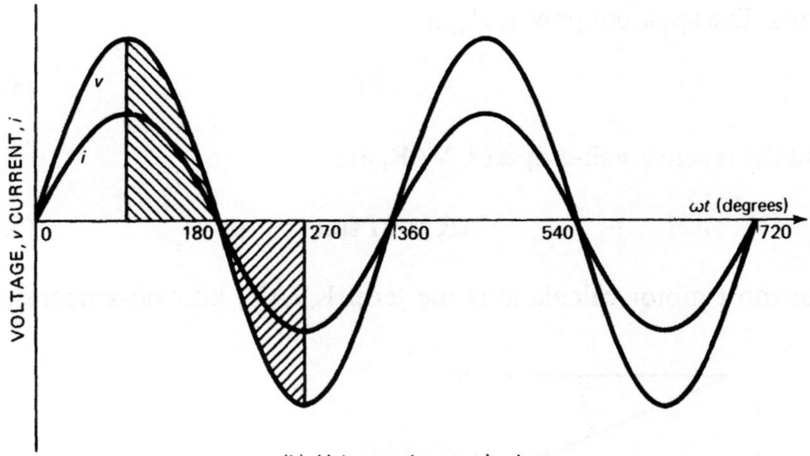


(b) Voltage and current in phase.

**Figure 5.2.** Voltage and current waveforms.

derivations will be omitted here, but the results are stated in the succeeding paragraphs.

For constant-frequency cases where the voltage and current waveforms are sinusoidal, phasors can be used as a convenient means to display the phase relationship between voltage and current. The phasor diagram representations of Figs. 5.1, 5.2(a), and 5.2(b) are shown in Fig. 5.3. The length of the phasor is proportional to the rms value of the voltage or current and the angle between $V$ and $I$ is the phase angle, $\theta$. Figure 5.3(a) is a general case in which the current lags behind the voltage by an angle $\theta$. Figure 5.3(b) and (c) are for $\theta = 90°$ and $\theta = 0°$, respectively. The average value of real power, $P$, is calculated using the equation

$$P = VI\cos\theta \tag{5.2}$$

where $V$ = rms value of voltage, $I$ = rms value of current, and $\theta$ = angle of phase shift between $V$ and $I$. The term $\cos\theta$ is called the power factor and it can range between $-1.0$ and $+1.0$. For motoring operation, the values of power factor are between 0 and 1.0. A negative value of power factor signifies induction generator operation. Usually, power factor, like efficiency, is expressed as a percentage value, for example, 90%, rather than the per-unit value, 0.9.

It is useful to define two additional terms for subsequent discussions. The apparent power, $P_A$, is

$$P_A = VI \tag{5.3}$$

and the reactive volt-amperes, VAR, are

$$\mathrm{VAR} = VI\sin\theta \tag{5.4}$$

For most motor calculations the term KVAR, kilovolt-amperes, is

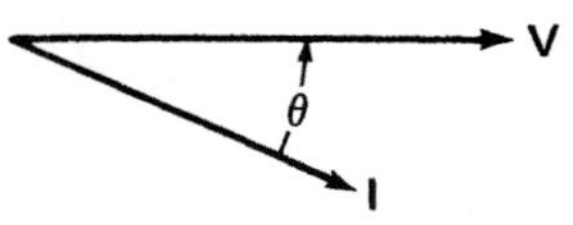


(a) Voltage and current phasors for Figure 5.1.

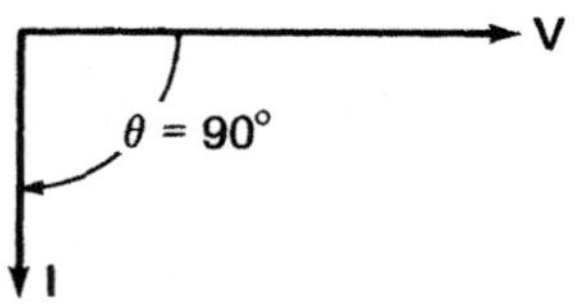


(b) Voltage and current phasors for Figure 5.2(a).

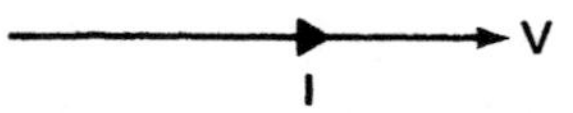


(c) Voltage and current phasors for Figure 5.2(b).

**Figure 5.3.** Voltage and current phasor diagrams.

more convenient to use. The relationship between the three quantities defined by (5.2), (5.3), and (5.4) can be displayed in a phasor diagram shown in Fig. 5.4.

Now, having defined the various components, let us consider their effects on energy consumption. The real power, as previously noted, is delivered to the motor as electrical power and is converted by the motor to mechanical power which is delivered to the load plus losses which escape as heat. This power has to be supplied by the power system and, therefore, some form of fuel—oil, coal, gas, nuclear, etc.—is consumed in its generation. The reactive volt-amperes provide the machine's magnetic fields and; as such, do not consume power from the power system. However, this does not mean that there is not a cost associated with providing these reactive volt-amperes. Indeed, there is a cost, and most power companies make a charge for this reactive energy.

The KVAR require current in the power company's transmission lines and this current results in $I^2r$ losses which are real power and do, therefore, consume fuel. Second, the size of both generators and transformers is determined by the current that they must provide, so that the initial capital investment in equipment is determined by the kilovolt-amperes that the equipment must provide, as well as by the power requirements of the system. Therefore, nearly all power companies make some kind of charge to their industrial users for the effects of power factor. Rate structures and the methods used to charge for reactive energy vary considerably among power companies. These costs are not so simply stated as the cost for the real power component of energy usage which can usually be stated as dollars per kilowatt-hour. It is also true that the real power costs for operating motors will exceed the cost associated with power factor by several times but, nevertheless, the cost of reactive energy is not insignificant in many cases and should be evaluated in making an analysis of energy usage.

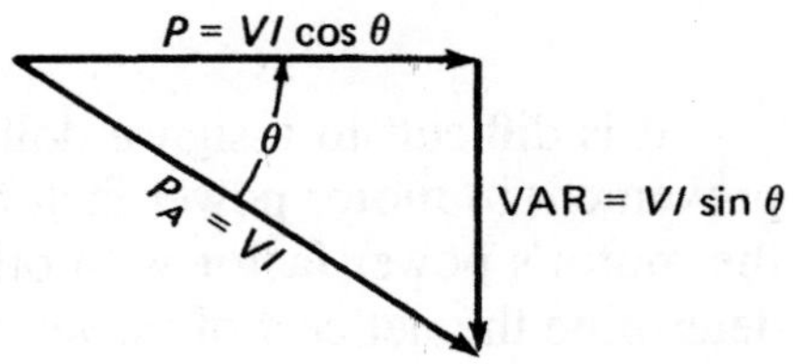


**Figure 5.4.** Phasor diagrams for real power, apparent power, and reactive volt-amperes.

## 5.2. Power Factor Penalty Cost Example

As an example to illustrate the costs associated with power factor, consider a typical industrial plant. Its electric bill consists of three separate charges.

1. Energy cost—a charge based on the total kilowatt-hours used
2. Demand charge—a charge for the peak kilowatt demand used at any time during the billing period
3. Power factor penalty—an escalation of the demand charge if the power factor falls below some specified minimum value

$$\text{Billing demand} = \frac{\begin{pmatrix}\text{Peak kilowatt}\\ \text{demand}\end{pmatrix} \times \begin{pmatrix}\text{Specified minimum}\\ \text{power factor}\end{pmatrix}}{(\text{Actual power factor})} \times \begin{pmatrix}\text{Demand}\\ \text{charge rate}\end{pmatrix}$$

The plant in one billing period used 240,000 kWh, its peak demand was 1700 kW, and the plant power factor was 70%. The applicable billing rates and conditions imposed by the electric company were:

1. Energy cost = \$0.05 per kWh
2. Demand charge = \$5.50 per kW demand
3. Minimum power factor to avoid a power factor penalty = 85%

The charges for the billing period are calculated as follows:

| | |
|---|---|
| Energy cost 240,000 × \$0.05 | = \$12,000.00 |
| Demand charge 1700 × \$5.50 | = 9,350.00 |
| Power factor penalty $\left(\frac{0.85}{0.70} - 1\right) \times 1700 \times \$5.50$ | = 2,003.57 |
| | \$23,353.57 |

It is difficult to assign a dollar value to a percentage point improvement in motor power factor because of the interdependence of the motor's power factor with other reactive loads in the plant that determine the net cost of power factor. In fact, it is entirely possible that the variation in other reactive loads in a plant or building may result in the reactive energy taken by motors having a much different dollar value from one month to the next.

One quantitative method of assigning a specific value to power factor differences among motors is to determine the cost of equipping the motors with sufficient power factor correcting capacitors to bring them up to some specified power factor. For this method of evaluation, some power factor needs to be determined to which all of the motors in the facility are going to be corrected. This decision is usually made based on the electrical requirements of the entire plant or building and the economics associated with operating the facility at an improved power factor. Then, for any given motor, the cost penalty associated with its power factor is the cost of purchasing capacitors to raise the motor's power factor to the specified level for the plant or building. Power factor correction and means for calculating the value of capacitance needed are the subjects of the next section.

## 5.3. Power Factor Correction

When an ac voltage is applied to a circuit containing a resistance and capacitance, energy is alternately stored in and discharged from an electrostatic field rather than a magnetic field as is true for an RL circuit. The current in the RC circuit leads the voltage as shown in Fig. 5.5. The phasor diagram showing the relationship between real

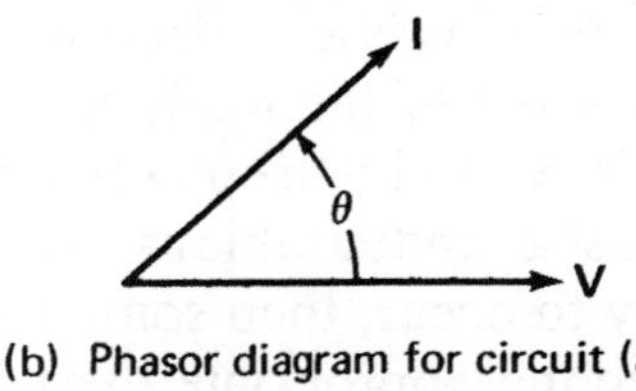


(a) Electric circuit with reactance and capacitance.

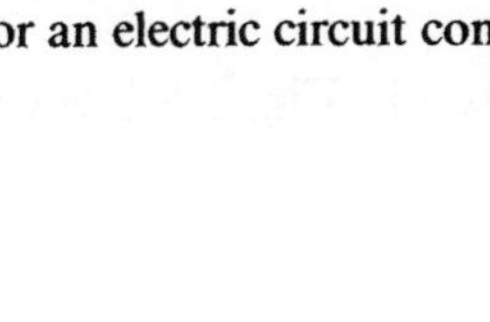


(b) Phasor diagram for circuit (a).

**Figure 5.5.** Phasor diagram for an electric circuit containing resistance and capacitance.

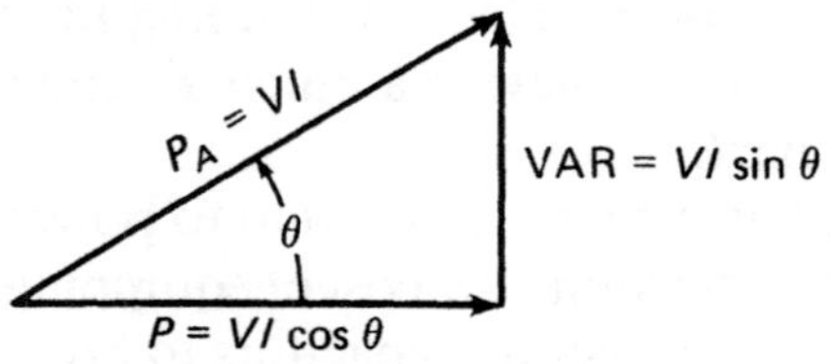


**Figure 5.6.** Phasor diagram for real power, apparent power, and reactive volt-amperes in an RC circuit.

power and reactive volt-amperes for an RC circuit is displayed in Fig. 5.6; this can be compared against Fig. 5.4 which shows the relationship of phasors for an RL circuit.

Now, if an inductance and capacitance are connected in parallel, the reactive volt-amperes can circulate between them, and fewer volt-amperes will have to be supplied by the power source. This is the idea behind power factor correction. Capacitors are connected in parallel with a motor and the apparent power factor of the motor–capacitor unit is increased because fewer volt-amperes are required from the line.

Two methods are used for connecting the power factor correcting capacitors in the system. The first method, shown in Fig. 5.7, is to connect a single bank of capacitors to the power distribution line and have it provide power factor correction for all of the loads on the line. The second method, shown in Fig. 5.8, is to connect the capacitors at or near the motor terminals and switch both the capacitors and the motor as a unit.

There are advantages to both systems. Connecting a single bank of capacitors to supply the entire distribution system, Fig. 5.7, reduces the number of capacitor installations in a multimotor system. The National Electrical Code[2] requires that, if this connection system is used, a disconnecting means must be installed to permit the capacitors to be removed as a regular operating procedure.

Overcorrection can be a problem with this type of installation. If the capacitors are selected to provide correction to a near-unity power factor when all of the system loads are operating, then the system may be overcorrected when some of the loads are removed and leading current will have to be delivered by the power supply. Too much leading current demand is just as undesirable as too much lagging current. If this situation is likely to occur, then some method of monitoring the load and disconnecting appropriate capacitor units should be provided.

The connection shown in Fig. 5.8 has the advantage of providing

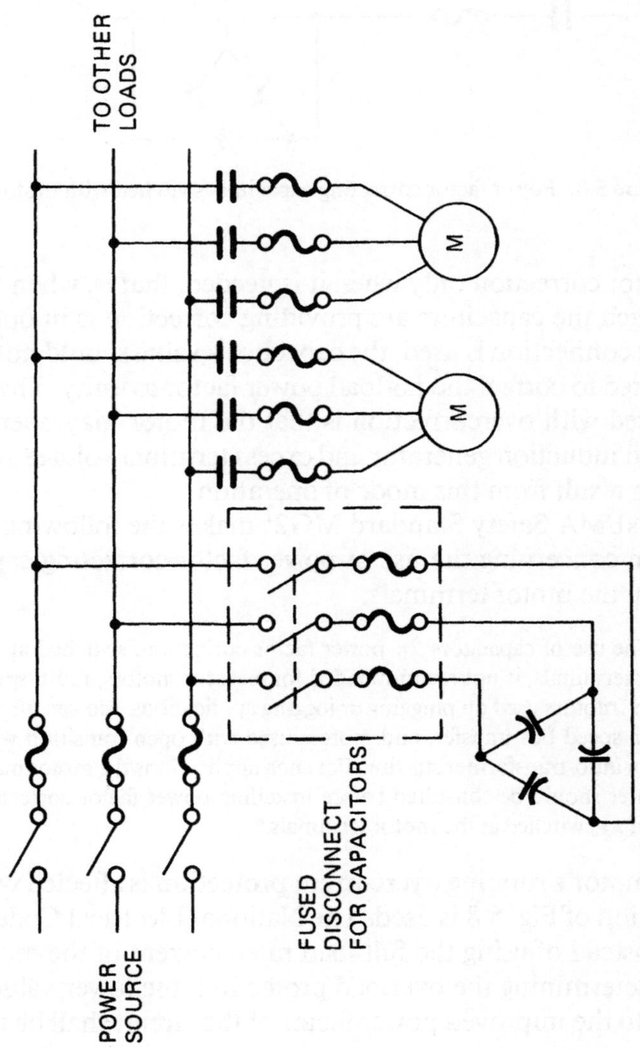


**Figure 5.7.** Power factor correcting capacitors installed as a single unit serving the distribution line.

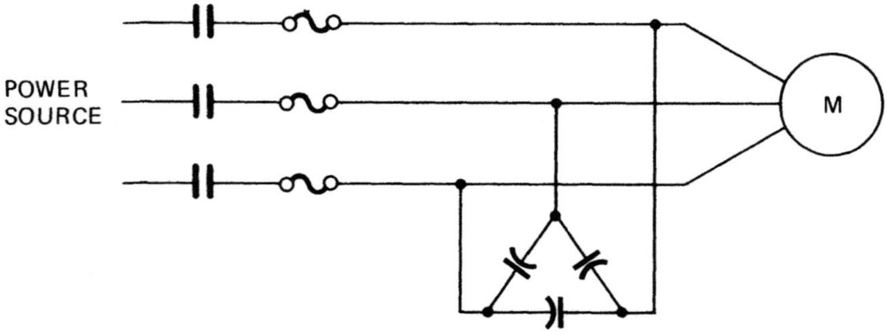


**Figure 5.8.** Power factor correcting capacitors switched with motor.

power factor correction only when it is needed, that is, when the motor for which the capacitors are providing correction is in operation. When this connection is used, the capacitor's value should not exceed that required to correct the no-load power factor to unity. The danger encountered with overcorrection is that the motor may operate as a self-excited induction generator and excess terminal voltage and shaft torque can result from this mode of operation.

The NEMA Safety Standard MG2[1] makes the following recommendation concerning the use of power factor correcting capacitors switched at the motor terminals:

> The use of capacitors for power factor correction, switched at the motor terminals, is not recommended for elevator motors, multi-speed motors, motors used on plugging or jogging applications, motors subject to high speed bus transfer, and motors used with open transition wye-delta or auto-transformer starting. For such applications the motor manufacturer should be consulted before installing power factor corrective capacitors switched at the motor terminals.*

The motor's running overcurrent protection is affected when the configuration of Fig. 5.8 is used. The National Electrical Code[2] specifies that instead of using the full-load rated current of the motor as a basis for determining the overload protection, the lower value corresponding to the improved power factor of the circuit shall be used.

* This paragraph is 3.16.10 of NEMA Standards Publication MG2, 1989. It is reproduced by permission of the National Electrical Manufacturers Association which holds the copyright.

## 5.4. Calculation of Power Factor Correction KVAC

The KVAC required to correct a motor's power factor from its uncorrected value, $PF_1$, to a new value, $PF_2$, is determined by finding the difference in KVARs required for the two conditions and then providing this difference in capacitive KVAC. The method can be visualized from the phasor diagram in Fig. 5.9 and the equations for making the calculations are

$$PF_1 = \cos\theta_1 \tag{5.5}$$

$$PF_2 = \cos\theta_2 \tag{5.6}$$

where $PF_1$ and $PF_2$ are the motor power factor uncorrected and the power factor of the motor and capacitor combination after correction, respectively.

$$P = \frac{\text{hp} \times 0.746}{\eta} \tag{5.7}$$

where hp = horsepower at which power factor correction is calculated, usually rated horsepower, and $\eta$ = motor efficiency.

Then the angles $\theta_1$ and $\theta_2$ can be calculated from (5.5) and (5.6)

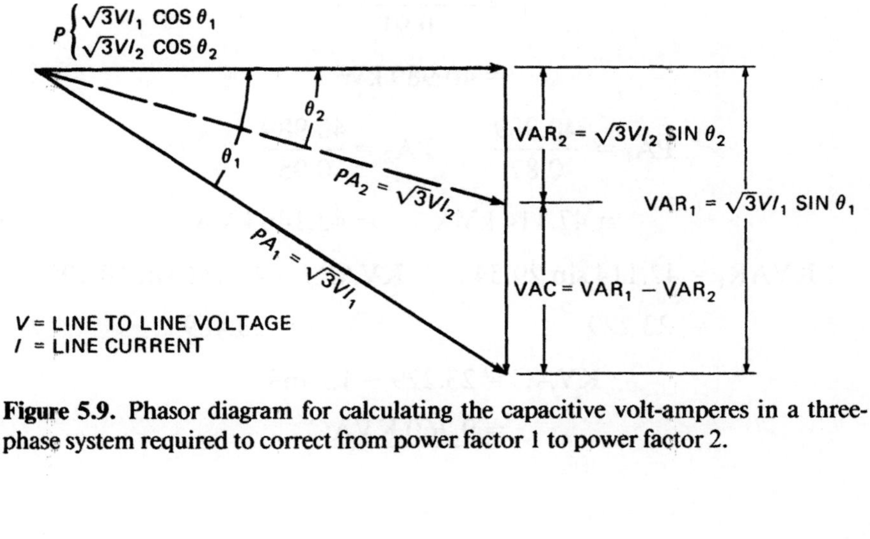


**Figure 5.9.** Phasor diagram for calculating the capacitive volt-amperes in a three-phase system required to correct from power factor 1 to power factor 2.

and the apparent powers $PA_1$ and $PA_2$, and $VAR_1$ and $VAR_2$ can be calculated using

$$PA_1 = \frac{P}{\cos\theta_1} \tag{5.8}$$

$$PA_2 = \frac{P}{\cos\theta_2} \tag{5.9}$$

$$VAR_1 = PA_1 \sin\theta_1 \tag{5.10}$$

$$VAR_2 = PA_2 \sin\theta_2 \tag{5.11}$$

Finally, the capacitive correction expressed as VAC is

$$VAC = VAR_1 - VAR_2 \tag{5.12}$$

EXAMPLE KVAC CALCULATION. It is desired to correct a 50-hp motor with a power factor of 87% up to a 95% under rated load. The motor's efficiency at rated load is 91%.

$$\theta_1 = \cos^{-1}0.87 \qquad \theta_2 = \cos^{-1}0.95$$

$$= 29.54° \qquad = 18.19°$$

$$P = \frac{50 \times 0.746}{0.91}$$

$$= 40.989 \text{ kW}$$

$$PA_1 = \frac{40.989}{0.87} \qquad PA_2 = \frac{40.989}{0.95}$$

$$= 47.114 \text{ kVA} \qquad = 43.146 \text{ kVA}$$

$$KVAR_1 = 47.114 \sin 29.54° \qquad KVAR_2 = 43.146 \sin 18.19°$$

$$= 23.229 \qquad = 13.469$$

$$KVAC = 23.229 - 13.469$$

$$= \underline{9.760 \text{ KVAC}}$$

An alternate method for calculating the KVAC necessary to raise a motor's power factor from one value to another is to use the kW multipliers shown in Table 5.1. An example of the use of this table follows.

EXAMPLE. Total kW input to the load is 100 kW. The power factor is 60%. The leading reactive KVAR necessary to raise the power factor to 90% is found by multiplying the 100 kW by the factor found in the table which is 0.850. Then 100 kW × 0.850 = 85.0 KVAR.

## 5.5. Self-Excitation of Induction Machines

This book and many other discussions of power-factor correction caution against overcorrecting. This caution stems from the fact that an induction machine can become self-excited and operate as an induction generator if certain conditions are fulfilled. Since the reference to this mode of operation is encountered so frequently in the power-factor correction literature, a brief explanation of the fundamentals of self-excitation is included here.

Induction motors are usually not thought of as also being generators, but they can be under certain conditions. If the motor is driven at a speed above its synchronous speed and is connected to a power line which can supply the required leading, exciting current, then it will, indeed, deliver electrical power into the power system. This method of power generation is not frequently used, as it suffers from several disadvantages, two major ones being that the machine must be connected to a power line to provide its excitation and it has poor inherent regulation characteristics compared with most synchronous generators.

This ability of an induction motor to act as a generator gives rise to the concern about overexcitation. If the power-factor capacitors remain connected to the motor terminals after it is disconnected from the power line, and if they are of a value such that the lagging volt-amperes of excitation equal the leading volt-amperes of the capacitors, then the machine will self-excite. Of course, the machine must be driven by an external source, but frequently, the inertia of the load is high enough to maintain the rotational speed of the motor long enough after disconnection from the power line for self-excitation to take place.

## Table 5.1. kW Multipliers for Determining Capacitor Kilovars[a]

| Original power factor in percent | Desired power factor in percent: 80 | 81 | 82 | 83 | 84 | 85 | 86 | 87 | 88 | 89 | 90 | 91 | 92 | 93 | 94 | 95 | 96 | 97 | 98 | 99 | 100 |
|---|---|---|---|---|---|---|---|---|---|---|---|---|---|---|---|---|---|---|---|---|---|
| 51 | .937 | .962 | .989 | 1.015 | 1.041 | 1.067 | 1.094 | 1.120 | 1.147 | 1.175 | 1.203 | 1.231 | 1.261 | 1.292 | 1.324 | 1.358 | 1.395 | 1.436 | 1.484 | 1.544 | 1.687 |
| 52 | .893 | .919 | .945 | .971 | .997 | 1.023 | 1.050 | 1.076 | 1.103 | 1.131 | 1.159 | 1.187 | 1.217 | 1.248 | 1.280 | 1.314 | 1.351 | 1.392 | 1.440 | 1.500 | 1.643 |
| 53 | .850 | .876 | .902 | .928 | .954 | .980 | 1.007 | 1.033 | 1.060 | 1.088 | 1.116 | 1.144 | 1.174 | 1.205 | 1.237 | 1.271 | 1.308 | 1.349 | 1.397 | 1.457 | 1.600 |
| 54 | .809 | .935 | .861 | .837 | .913 | .939 | .966 | .992 | 1.019 | 1.047 | 1.075 | 1.103 | 1.133 | 1.164 | 1.196 | 1.230 | 1.267 | 1.308 | 1.356 | 1.416 | 1.559 |
| 55 | .769 | .795 | .821 | .847 | .873 | .899 | .926 | .952 | .979 | 1.007 | 1.035 | 1.063 | 1.090 | 1.124 | 1.156 | 1.190 | 1.228 | 1.268 | 1.316 | 1.377 | 1.519 |
| 56 | .730 | .756 | .782 | .808 | .834 | .860 | .887 | .913 | .940 | .968 | .996 | 1.024 | 1.051 | 1.085 | 1.117 | 1.151 | 1.189 | 1.229 | 1.277 | 1.338 | 1.480 |
| 57 | .692 | .718 | .744 | .770 | .796 | .822 | .849 | .875 | .902 | .930 | .958 | .986 | 1.013 | 1.047 | 1.079 | 1.113 | 1.151 | 1.191 | 1.239 | 1.300 | 1.442 |
| 58 | .655 | .681 | .707 | .733 | .759 | .785 | .812 | .838 | .865 | .893 | .921 | .949 | .976 | 1.010 | 1.042 | 1.076 | 1.114 | 1.154 | 1.202 | 1.263 | 1.405 |
| 59 | .618 | .644 | .670 | .696 | .722 | .748 | .775 | .801 | .828 | .856 | .884 | .912 | .939 | .973 | 1.005 | 1.039 | 1.077 | 1.117 | 1.165 | 1.226 | 1.368 |
| 60 | .584 | .610 | .636 | .662 | .688 | .714 | .741 | .767 | .794 | .822 | .850 | .878 | .905 | .939 | .971 | 1.005 | 1.043 | 1.083 | 1.131 | 1.192 | 1.334 |
| 61 | .549 | .575 | .601 | .627 | .653 | .679 | .706 | .732 | .759 | .787 | .815 | .843 | .870 | .904 | .936 | .970 | 1.008 | 1.048 | 1.096 | 1.157 | 1.299 |
| 62 | .515 | .541 | .567 | .593 | .619 | .645 | .672 | .698 | .725 | .753 | .781 | .809 | .836 | .870 | .902 | .936 | .974 | 1.014 | 1.062 | 1.123 | 1.265 |
| 63 | .483 | .509 | .535 | .561 | .587 | .613 | .640 | .666 | .693 | .721 | .749 | .777 | .804 | .838 | .870 | .904 | .942 | .982 | 1.030 | 1.091 | 1.233 |
| 64 | .450 | .476 | .502 | .528 | .544 | .580 | .607 | .633 | .660 | .688 | .716 | .744 | .771 | .805 | .837 | .871 | .909 | .949 | .997 | 1.058 | 1.200 |
| 65 | .419 | .445 | .471 | .497 | .523 | .549 | .576 | .602 | .629 | .657 | .685 | .713 | .740 | .774 | .806 | .840 | .878 | .918 | .966 | 1.027 | 1.169 |
| 66 | .398 | .414 | .440 | .466 | .492 | .518 | .545 | .571 | .598 | .626 | .654 | .682 | .709 | .743 | .775 | .809 | .847 | .887 | .935 | .996 | 1.138 |
| 67 | .358 | .384 | .410 | .436 | .462 | .488 | .515 | .541 | .568 | .596 | .624 | .652 | .679 | .713 | .745 | .779 | .817 | .857 | .905 | .966 | 1.108 |
| 68 | .329 | .355 | .381 | .407 | .433 | .459 | .486 | .512 | .539 | .567 | .595 | .623 | .650 | .684 | .716 | .750 | .788 | .828 | .876 | .937 | 1.079 |
| 69 | .299 | .325 | .351 | .377 | .403 | .429 | .456 | .482 | .509 | .537 | .565 | .593 | .620 | .654 | .686 | .720 | .758 | .798 | .840 | .907 | 1.049 |
| 70 | .270 | .296 | .322 | .348 | .374 | .400 | .427 | .453 | .480 | .508 | .536 | .564 | .591 | .625 | .657 | .691 | .729 | .769 | .811 | .878 | 1.020 |
| 71 | .242 | .268 | .294 | .320 | .346 | .372 | .399 | .425 | .452 | .480 | .508 | .536 | .563 | .597 | .629 | .663 | .701 | .741 | .783 | .850 | .992 |
| 72 | .213 | .239 | .265 | .291 | .317 | .343 | .370 | .396 | .423 | .451 | .479 | .507 | .534 | .568 | .600 | .634 | .672 | .712 | .754 | .821 | .963 |
| 73 | .186 | .212 | .238 | .264 | .290 | .316 | .343 | .369 | .396 | .424 | .452 | .480 | .507 | .541 | .573 | .607 | .645 | .685 | .727 | .794 | .936 |

Original power factor in percent

| | | | | | | | | | | | | | | | | | | | | | |
|---|---|---|---|---|---|---|---|---|---|---|---|---|---|---|---|---|---|---|---|---|---|
| 74 | .159 | .185 | .211 | .237 | .263 | .289 | .316 | .342 | .369 | .397 | .425 | .453 | .480 | .514 | .546 | .580 | .618 | .658 | .700 | .767 | .909 |
| 75 | .132 | .158 | .184 | .210 | .236 | .262 | .289 | .315 | .342 | .370 | .398 | .426 | .453 | .487 | .519 | .553 | .591 | .631 | .673 | .740 | .882 |
| 76 | .105 | .131 | .157 | .183 | .209 | .235 | .262 | .288 | .315 | .343 | .371 | .399 | .426 | .460 | .492 | .526 | .564 | .604 | .652 | .713 | .855 |
| 77 | .079 | .105 | .131 | .157 | .183 | .209 | .236 | .262 | .289 | .317 | .345 | .373 | .400 | .434 | .466 | .500 | .538 | .578 | .620 | .687 | .829 |
| 78 | .053 | .079 | .105 | .131 | .157 | .183 | .209 | .236 | .263 | .291 | .319 | .347 | .374 | .408 | .440 | .474 | .512 | .552 | .594 | .661 | .803 |
| 79 | .026 | .052 | .078 | .104 | .130 | .156 | .183 | .209 | .236 | .264 | .292 | .320 | .347 | .381 | .413 | .447 | .485 | .525 | .567 | .634 | .776 |
| 80 | .000 | .026 | .052 | .078 | .104 | .130 | .157 | .183 | .210 | .238 | .266 | .294 | .321 | .355 | .387 | .421 | .459 | .499 | .541 | .608 | .750 |
| 81 | — | .000 | .026 | .052 | .078 | .104 | .131 | .157 | .184 | .212 | .240 | .268 | .295 | .329 | .361 | .395 | .433 | .473 | .515 | .582 | .724 |
| 82 | — | — | .000 | .026 | .052 | .078 | .105 | .131 | .158 | .186 | .214 | .242 | .269 | .303 | .335 | .369 | .407 | .447 | .489 | .556 | .698 |
| 83 | — | — | — | .000 | .026 | .052 | .079 | .105 | .132 | .160 | .188 | .216 | .243 | .277 | .309 | .343 | .381 | .421 | .463 | .530 | .672 |
| 84 | — | — | — | — | .000 | .026 | .053 | .079 | .106 | .134 | .162 | .190 | .217 | .251 | .283 | .317 | .355 | .395 | .437 | .504 | .645 |
| 85 | — | — | — | — | — | .000 | .027 | .053 | .080 | .108 | .136 | .164 | .191 | .225 | .257 | .291 | .329 | .369 | .417 | .478 | .620 |
| 86 | — | — | — | — | — | — | — | .026 | .053 | .081 | .109 | .137 | .167 | .198 | .230 | .265 | .301 | .343 | .390 | .451 | .593 |
| 87 | — | — | — | — | — | — | — | — | .027 | .055 | .082 | .111 | .141 | .172 | .204 | .238 | .275 | .317 | .364 | .425 | .567 |
| 88 | — | — | — | — | — | — | — | — | — | .028 | .056 | .084 | .114 | .145 | .177 | .211 | .248 | .290 | .337 | .398 | .540 |
| 89 | — | — | — | — | — | — | — | — | — | — | .028 | .056 | .086 | .117 | .149 | .183 | .220 | .262 | .309 | .370 | .512 |
| 90 | — | — | — | — | — | — | — | — | — | — | — | .028 | .058 | .089 | .121 | .155 | .192 | .234 | .281 | .342 | .484 |
| 91 | — | — | — | — | — | — | — | — | — | — | — | — | .030 | .061 | .093 | .127 | .164 | .206 | .253 | .314 | .456 |
| 92 | — | — | — | — | — | — | — | — | — | — | — | — | — | .031 | .063 | .097 | .134 | .176 | .223 | .284 | .426 |
| 93 | — | — | — | — | — | — | — | — | — | — | — | — | — | — | .032 | .066 | .103 | .145 | .192 | .253 | .395 |
| 94 | — | — | — | — | — | — | — | — | — | — | — | — | — | — | — | .034 | .071 | .113 | .160 | .221 | .363 |
| 95 | — | — | — | — | — | — | — | — | — | — | — | — | — | — | — | — | .037 | .079 | .126 | .187 | .328 |

[a] Table reproduced by permission of Aerovox, North Dartmouth, Massachusetts. Predecimal zeros have been omitted.

The situation can be visualized by use of the equivalent circuit suggested by Wagner[3] and shown in Fig. 5.10. The no-load case, Fig. 5.10(b), represents the limiting case under which self-excitation will occur with the smallest value of capacitive KVAC.[3] For no load, if self-excitation is to be maintained, the reactive kilovolt-amperes must sum to zero which means $x_1 + X_m$ must equal $X_c$. However, $X_m$ is really not constant but varies with saturation as the terminal voltage varies, which it is free to do once the motor is disconnected from the line. Figure 5.11 displays a typical no-load saturation curve.

Values of $X_c$ corresponding to capacitance applied to the motor terminals can be plotted on the same graph with the saturation curve as shown in Fig. 5.11. The line labeled $X_{c1}$ is for one typical value of capacitance. In this case $X_{c1}$ intersects the motor saturation curve and the machine would self-excite. As the value of capacitance decreases, the slope of the $X_c$ lines increase until, for some value of capacitance, the $X_c$ line will coincide with the air gap line. At this point, the machine will no longer self-excite and any smaller value of capacitance may safely be used without the danger of self-excitation. The situation is somewhat analogous to the shunt excited dc generator,[5] which has a limiting value of field circuit resistance above which self-excitation will not take place.

The danger in self-excitation is that with the power system disconnected from the motor terminals, the motor–capacitor combination no longer has any regulating means to control the generated volt-

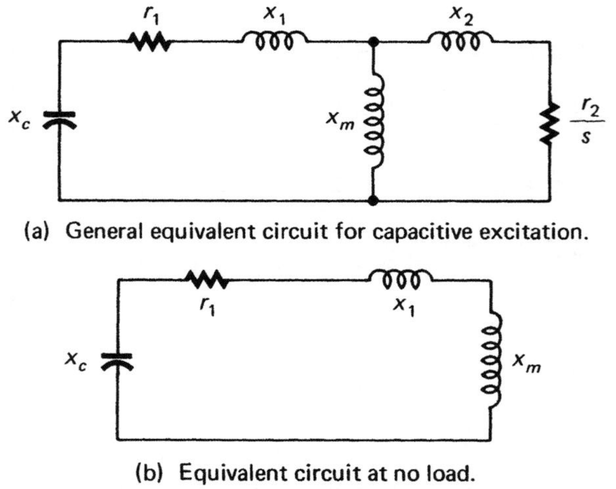


**Figure 5.10.** Equivalent circuit of induction generator with capacitor excitation.

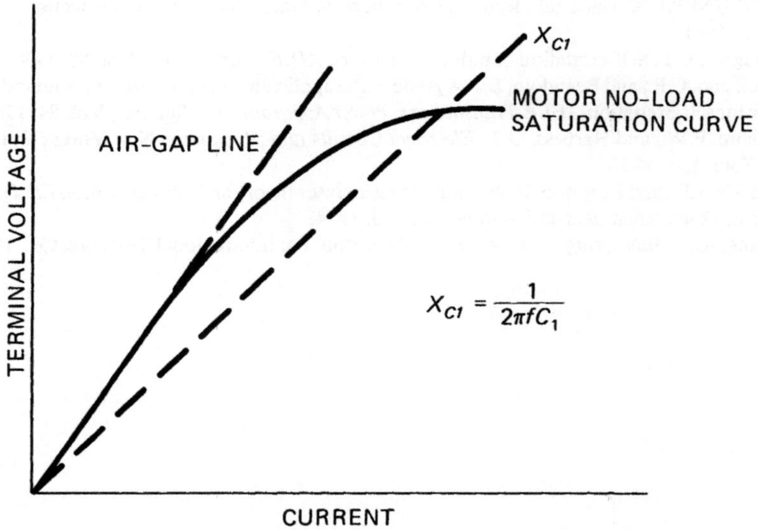


**Figure 5.11.** Induction motor saturation curve and $X_c$ for self-excitation.

age and the terminal voltage may rise to dangerously high values. This period of unregulated self-excitation may also be accompanied by severe torque transients which could damage the motor shaft or driven equipment.

This discussion illustrates in general terms the principles involved in the self-excitation. The situation which actually occurs in practice is a sudden removal of power to the motor, the motor starts to slow down, the self-excitation builds up, and a braking torque is developed. This is essentially a transient phenomenon and the application of a steady-state equivalent circuit to this situation is not completely valid but it does aid in a physical insight into the problem. References 4 and 6 both deal with the self-excitation problem in more rigorous detail. However, this rigorous analysis is fortunately not necessary for most power factor correcting application problems. The general guideline stated previously of limiting the correcting KVAC to a value equal to or less than that required to raise the no-load power factor to unity is a simple rule sufficient for most applications.

## References

1. NEMA Standards Publication No. MG2, "Safety Standard for Construction and Guide for Selection, Installation and Use of Electric Motors and Generators." Washington, D.C.: National Electrical Manufacturers Association, 1989.

2. ANSI/NFPA 70, National Electrical Code. Boston, Mass.: National Fire Protection Association, 1981.
3. Wagner, C.F. Self-excitation of induction motors, *AIEE Transactions,* Vol. 58, 1939.
4. DeSieno, C.F., and Beaudoin, B.J. A guide to the application of capacitors without induction motor self-excitation, *IEEE Transactions, Power Apparatus and Systems,* Vol. 84, 1965.
5. Hehre, F.W., and Harness, G.T. *Electrical Circuits and Machinery.* New York: John Wiley & Sons, Inc., 1944.
6. Barkle, J.E., and Ferguson, R.W. Induction generator theory and application, *AIEE Transactions, Power Apparatus and Systems,* Vol. 73, 1954.
7. Tang, K.Y. *Alternating Current Circuits.* Scranton, Pa: International Textbook Co., 1940.

# 6

# Single-Phase and Synchronous Motors

Most of this book has been devoted to the medium-size, polyphase induction motors since this category is responsible for the major segment of the motor energy usage in the United States. However, energy saving opportunities also exist with other available motor types.

Below 1 hp, most motors are single-phase and are categorized as fractional-horsepower or subfractional-horsepower motors. In the very large horsepowers, starting at ratings 15,000 hp and greater, excited-field synchronous machines are usually used. There is a large range of horsepower sizes from a few hundred horsepower up to approximately 15,000 hp in which both induction and excited-field synchronous machines are applied depending on the motor speed and needs of the application, so precise boundaries are difficult to define. These two categories, fractional-horsepower motors and large synchronous machines, are discussed in this chapter. Also, the permanent-magnet synchronous motor is discussed in this chapter. Many of these motors fall into the fractional-horsepower category, but the availability of high-energy magnets has extended the range for permanent-magnet motors well into the integral horsepower sizes.

## 6.1. Single-Phase Motor Types

The number of different types of single-phase motors is myriad. The fractional-horsepower, single-phase universe includes both synchronous and induction machines. The most common types are described briefly below and some additional discussion is presented in later sections on the motor types for which energy-efficient versions are available.

The feature which most frequently distinguishes one type of single-phase motor from another is the method of starting. A single-phase motor, either synchronous or induction, does not inherently have any starting torque. Thus, some auxiliary means must be provided for starting the motor, and it is this auxiliary means that distinguishes one type of motor from another.

## Shaded-Pole Induction Motors

Shaded-pole induction machines are widely used in the subfractional horsepower range, 1/20th hp and below. The rotor consists of the squirrel-cage construction with bars short-circuited at the ends by short-circuiting rings. The stators are usually built with salient poles around which the main winding is wound. One segment of each main pole is surrounded by a short-circuited turn called the "shading coil." This shading coil causes a delay in flux buildup and decay for the region surrounded by the coil. Any change in flux through this short-circuited coil causes current flow and, therefore, an opposing mmf that delays the flux change relative to the flux in the main pole. The time displacement of flux in the shaded portion of the pole is sufficient to produce a form of rotating magnetic field in the air gap, and thus starting torque is produced.

## Hysteresis Motors

Hysteresis motors are synchronous motors. The rotor is a cylinder of hard magnetic materials and contains neither slots nor windings. The rotor is magnetized by the rotating magnetic field in the air gap, but because of hysteresis, the rotor's field lags behind the air-gap field and the necessary angle exists for torque production. So long as the rotor speed is less than synchronous, torque generation takes place in this manner. When the rotor reaches synchronous speed, the magnetized rotor poles lock into step with stator field and the motor operates as a synchronous machine. The idealized speed–torque curve

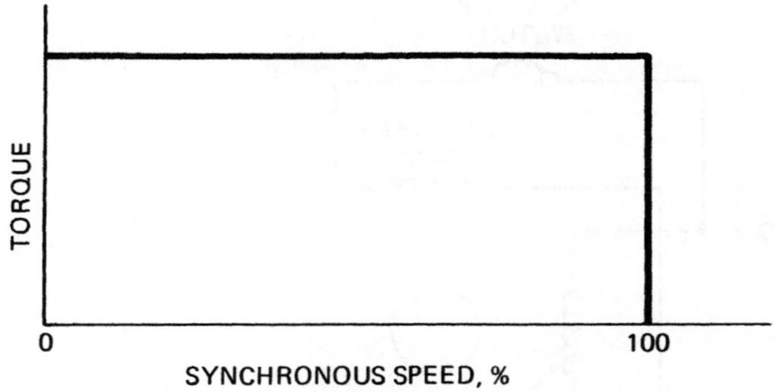


**Figure 6.1.** Idealized speed-torque curve of hysteresis motor.

is displayed in Fig. 6.1 and is unique in that the accelerating and synchronous torques are the same.

Hysteresis motors have primarily been used for timing devices and clock motors, in size ranges from a few watts up to fractional horsepower.

## Split-Phase Motors

Split-phase motors are single-phase induction motors and usually have a squirrel-cage rotor. The stator consists of two windings, namely, a main and auxiliary winding. Both are wound for the same number of poles but are displaced in space from one another by approximately 90 electrical degrees although, in some designs, other space displacements are used. The auxiliary winding is normally a higher resistance winding than the main winding, and the difference in winding impedance is responsible for the phase displacement between the fields established by the two windings. Thus, a rotating magnetic field is produced in the air gap and torque is produced by the action of the field on the squirrel-cage rotor. A starting switch which is usually centrifugally operated, disconnects the auxiliary winding after the motor has attained approximately 70 to 80% of synchronous speed. A diagram for a split-phase motor is displayed in Fig. 6.2.

Split-phase motors usually do not achieve the optimal 90° phase displacement between fields produced by the two windings and therefore are characterized by low starting torque. They are used in many types of appliances such as domestic refrigerators and home dryers.

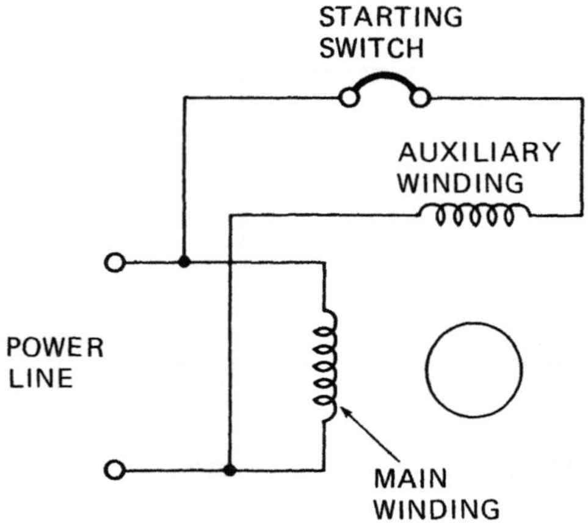


**Figure 6.2.** Split-phase motor wiring diagram.

Although split-phase motors are manufactured in large quantities, the intermittent service which characterizes many of their applications reduces the total energy consumption attributable to them.

## Capacitor-Start Motors

The capacitor-start variety is the versatile, general-purpose motor of the fractional-horsepower motor spectrum. Its rugged construction, low cost, and high starting torque capability make it useful for its wide range of applications. The capacitor-start motor, like the split-phase machine, has a squirrel-cage rotor and both a main and an auxiliary winding on the stator. A capacitor in series with the auxiliary winding provides the desired 90° displacement between the main and auxiliary fields, resulting in higher starting torque than is generally available in split-phase motors. The auxiliary winding and its capacitor are switched out of the circuit by a starting switch, which is usually centrifugally operated, as the motor nears synchronous speed. Electronic starting switches are also used in place of centrifugal switches.

Home workshops, industrial conveyors, pumps, and a plethora of other applications utilize capacitor-start motors. A typical wiring diagram for the capacitor-start motor is shown in Fig. 6.3.

## Two-Value Capacitor Motors

This motor starts as a capacitor-start motor having a capacitor in series with the auxiliary winding. As the motor nears synchronous

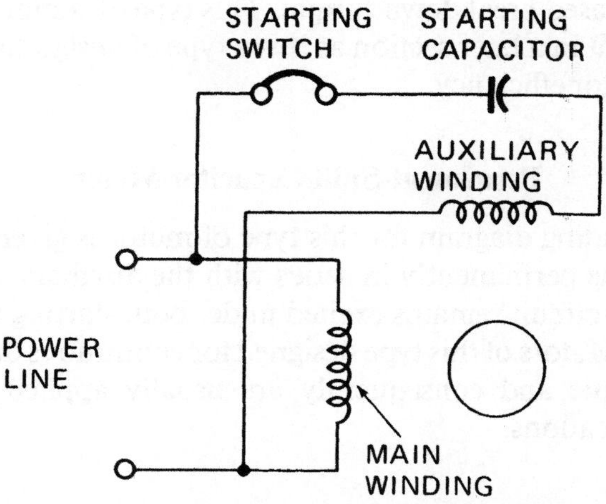


**Figure 6.3.** Capacitor-start motor wiring diagram.

speed, the starting capacitor is switched out but a running capacitor remains in the circuit and in series with the auxiliary winding. The wiring diagram is displayed in Fig. 6.4. Leaving the auxiliary winding and a series capacitor in the circuit for running conditions makes the motor operate more efficiently, with an improved power factor and

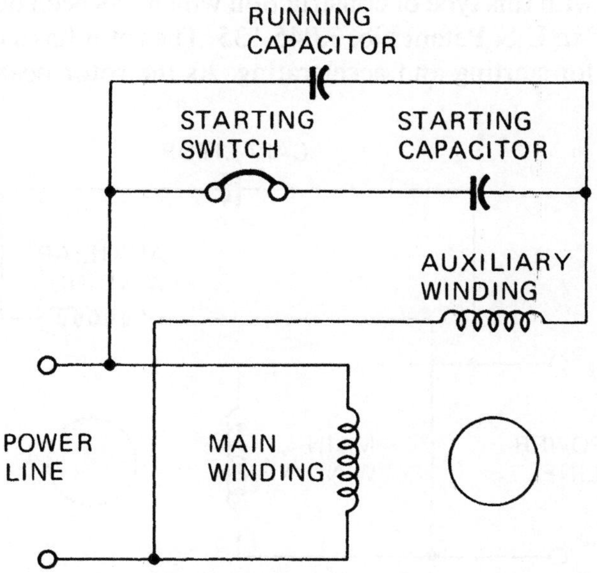


**Figure 6.4.** Two-value capacitor motor.

slightly increased breakdown torque. This type of motor is discussed in more detail in a later section as it is a type of design that is used to improve motor efficiency.

### Permanent-Split Capacitor Motor

The winding diagram for this type of motor is given in Fig. 6.5. A capacitor is permanently in series with the auxiliary winding and the auxiliary circuit remains excited under both starting and running conditions. Motors of this type designed for continuous duty have low starting torque and consequently are usually applied in fan and blower applications.

### Reluctance-Synchronous and Permanent-Magnet Synchronous Motors

These two types of synchronous motors are included in this section on fractional-horsepower motors because they are usually built in ratings 1 hp and below. Neither type of motor has an excited field as is used on larger synchronous machines. The reluctance-synchronous motor has a rotor structure with salient poles. Figure 6.6 displays one such rotor with this type of construction which has been described by Honsinger[5] in U.S. Patent No. 3,045,135. The rotor has a cast aluminum cage for starting and accelerating. As the rotor nears synchro-

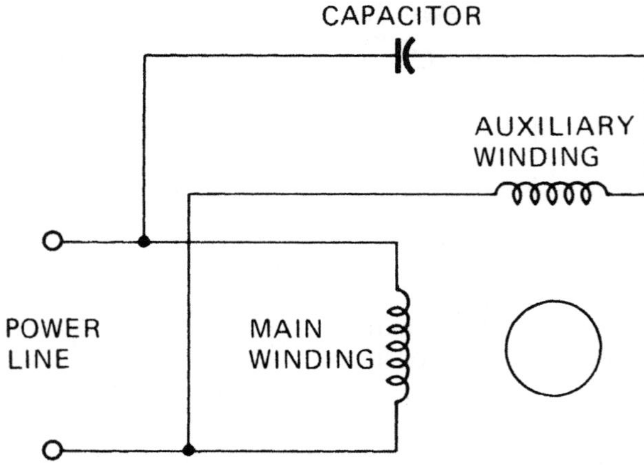


**Figure 6.5.** Permanent split capacitor motor.

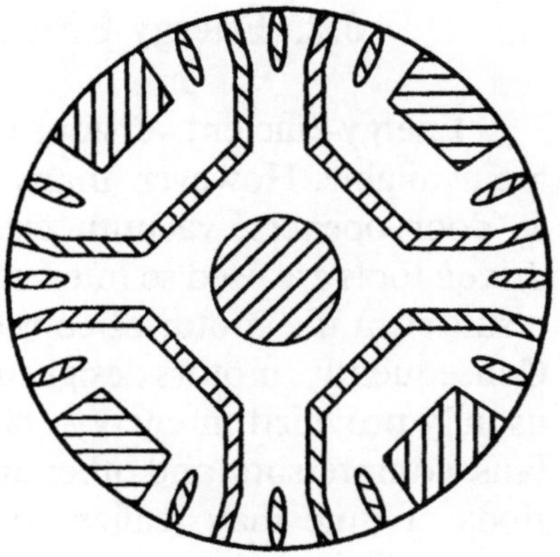

**Figure 6.6.** Cross-sectional view of a rotor for a reluctance-synchronous motor.

nous speed, it pulls into step as the salient poles "attach" themselves to the rotating magnetic field and the motor operates at synchronous speed.

The permanent-magnet synchronous motor has its permanent magnets embedded inside of the rotor structure and surrounded by a squirrel cage to provide starting torque. Once the rotor has pulled into step the machine operates as a synchronous motor similar to an excited-field type of synchronous machine. However, in this case, the field mmf is provided by the permanent magnets rather than through electromagnetic excitation. Further discussion of this type of motor is given in a later section at it is being applied to achieve energy savings, particularly in variable-speed applications.

## Universal Motors

The universal motor is a series wound motor which can be operated from either direct current or single-phase alternating current. The rotating member is wound like a dc armature and has a commutator and brushes. The stator member may be wound either as a distributed wound induction motor stator or as a salient pole stator similar to a dc machine. These motors are usually limited to ratings below 1 hp and often run at speeds of 10,000 rpm or higher, particularly at no load. Typical applications are vacuum cleaners, business machines, and home appliances.

References 1 and 2 provide complete and detailed descriptions of many types of single-phase motors.

## 6.2. Energy-Efficient, Single-Phase Motors

Energy-efficient versions of several types of single-phase motors are available. However, many single-phase motor applications such as door openers, vacuum cleaners, small appliances, and motor-driven tools are used so intermittently that little energy savings would result even if a motor of considerably higher efficiency was applied. Consequently, motors designed for these types of applications are not usually provided in energy-efficient models. Motors driving pumps, fans, compressors, and other applications which are used for long periods of time may realize considerable benefits from applying an energy-efficient design.

Split-phase, permanent-split capacitor and capacitor-start motors are all available in energy-efficient versions. Some of these designs are an extension of the techniques used for efficiency improvement discussed in earlier chapters of this book:

- Increased conductor area to reduce stator $I^2r$ loss
- Increased rotor conductor area to reduce rotor $I^2r$ loss
- Improved grade of steel and longer core length to reduce iron loss
- Improved ventilation to reduce windage losses and temperature rise

There is, however, one additional step which can be taken in a single-phase motor to improve its efficiency. This improvement consists of connecting a capacitor in series with the auxiliary winding for normal running operation. The reason this helps achieve the desired results is as follows:

A single-phase winding can be viewed as producing two equal and opposite rotating magnetic fields in the air gap. When the motor is at standstill, these two fields are of equal magnitude and produce equal and opposite torques so that the net starting torque is zero. Hence, a starting winding is used to modify the rotating fields so that they have a predominate component rotating in only one direction much like a polyphase motor. This provides the motor with a starting torque.

As the motor increases in speed, the torque in the forward direction increases up to the breakdown torque point and the backward torque diminishes, because of the demagnetizing effect of the currents induced in the rotor by the backward revolving field. A typical speed–torque curve of a single-phase motor with only the main winding excited is shown in Fig. 6.7. When the rotor runs at normal operating speed, approximately 0.02 to 0.04 slip, the backward field torque is small. However, the $I_{2b}^2 r_2$ losses related to currents induced in the rotor by the backward revolving field are not negligible. In one typical 3/4-hp design the $I_{2b}^2 r_2$ losses were 18% of the total losses.

A means of nullifying the effects of the backward field is to connect a capacitor in series with the auxiliary winding and leave both in the circuit during the run condition. Figure 6.4 displays a circuit which does this. The running capacitor and the auxiliary winding form a second phase for the otherwise single-phase motor. Auxiliary windings are normally displaced in the motor by 90 electrical degrees in space from the main winding, and, ideally, the capacitor would cause the current in the auxiliary winding to lead the current in the main winding by 90 electrical degrees, thus producing a two-phase motor. Since the effective impedance of the auxiliary winding is affected by motor slip, this ideal tuning will occur at only one load point and will be something less than ideal at other loads. Nevertheless, running capacitors are beneficial over a reasonable load range and are used in energy-efficient, single-phase motor designs. The running capacitor plus the other efficiency improvement techniques listed at the beginning of this section can result in loss reduction on some ratings of 40% and greater relative to the standard models of single-phase motors.

Running capacitors are added both to capacitor-start motors,

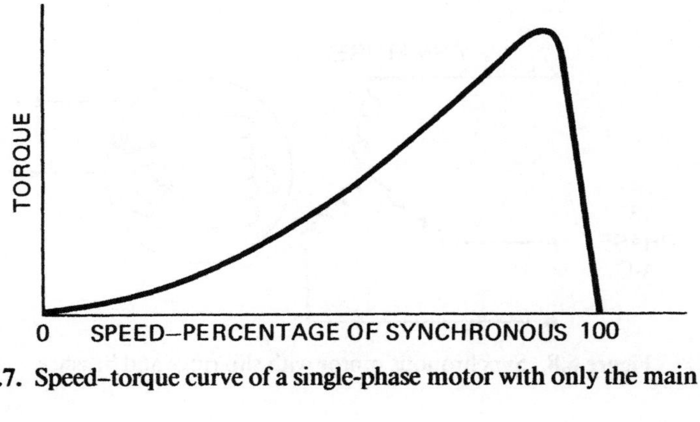


**Figure 6.7.** Speed–torque curve of a single-phase motor with only the main winding excited.

converting them to two-value capacitor motors, and to split-phase motors as means for improving operating efficiency.

## 6.3. Synchronous Motors, Excited-Field Type

Excited-field synchronous motors are not commonly used in the small and medium horsepower sizes but their frequency of application increases as the horsepower ratings increase. The performance characteristics which influence a synchronous motor's selection for a given application and, in particular, the motor's effect on energy savings are discussed in this section.

Synchronous motors of the type discussed here have an electrically excited field which rotates in synchronism with the rotating field of the stator winding. The stator winding is usually referred to as the armature winding in a synchronous motor. The field is normally on the rotating member and excited by dc either through brushes and slip rings or via a brushless exciter. Diagrams for the two types of excitation systems are shown in Figs. 6.8 and 6.9. In a slip ring motor, the field excitation is supplied from a dc source by conduction through the brushes and slip rings directly to the field. The brushless excitation system is comprised of an ac exciter winding, a rectifier to convert the ac to dc, and the motor field winding all mounted on the same shaft and contained inside of a single motor housing.

The separate source of excitation supplied to a synchronous motor provides it with a degree of flexibility not found in an induction machine. A synchronous motor's power factor can be controlled by

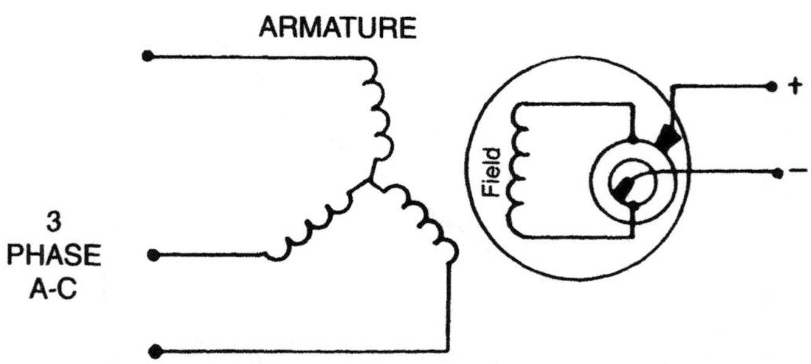


**Figure 6.8.** Synchronous motor with slip rings and brushes.

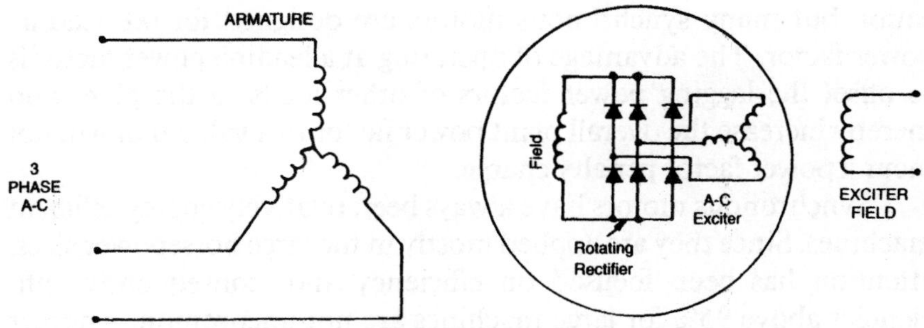


**Figure 6.9.** Excitation systems for a brushless synchronous motor.

controlling the field excitation. This characteristic is evident by examining the V curves, Fig. 6.10, which are plots of the line current versus field excitation. Figure 6.10 displays the V curves for three different conditions: full load, half load, and no load. In each case, the current reaches some minimum value as the excitation is increased and then increases again with further increase in field excitation. This minimum value of current corresponds to unity power factor operation. Segments of the curves lying to the right of the unity power factor point correspond to leading power factor operation. Synchronous motors are designed for operation at some specific power factor under full-load conditions. Unity power factor is a common design specifi-

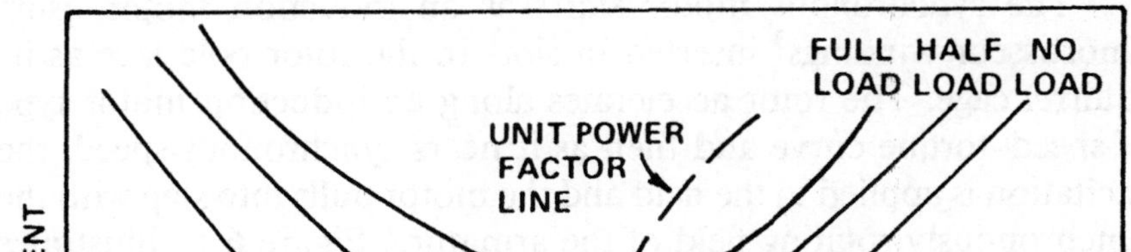


**Figure 6.10.** Synchronous motor V curves.

cation but many synchronous motors are designed for 0.8 leading power factor. The advantage of operating at a leading power factor is to offset the lagging power factors of other loads in the plant and thereby increase the overall plant power factor to a value that will not incur a power factor penalty charge.

Synchronous motors have always been relatively energy-efficient machines. Since they are applied mostly in the large horsepower sizes, attention has been focused on efficiency and, consequently, efficiencies above 95% for large machines are not uncommon. Another area dominated by synchronous motors is slow-speed applications. Slow speeds favor synchronous motors because they can be built with a larger air gap than the equivalent rating in an induction motor and the magnetization can be supplied by the dc field. Thus, the synchronous motor can be operated at unity or even leading power factor whereas a slow-speed induction motor may operate with a power factor in the 60s or 70s. Efficiency for the slow-speed, large-horsepower synchronous motor is usually better than the equivalent induction motor.

As in most engineering problems, there are trade-offs to consider when choosing between synchronous and induction motors. Starting a synchronous motor is more complex than starting an induction motor, and $Wk^2$ accelerating capabilities are often more limited for synchronous than for induction motors.

The synchronous motor starts as an induction motor using amortisseur windings[3] inserted in slots in the rotor pole face as its squirrel cage. The rotor accelerates along an induction motor type of speed–torque curve and then as it nears synchronous speed, the excitation is applied to the field and the motor pulls into step with the synchronously rotating field of the armature. Figure 6.11 illustrates the speed–torque curve of a synchronous motor. The "pull-in" torque is the value of load torque which the machine will synchronize and is dependent on the $Wk^2$ of the load. The "pull-out" torque is the value of load torque which, if exceeded, will cause the machine to pull out of step and operate back on the induction motor speed–torque curve. Continuous operation at loads above pull-out torque cannot be sustained, as the motor vibrates excessively and will overheat.

During starting, the motor field is shorted, usually through a resistor, to avoid high voltage being induced at the field terminals. Then, as the motor nears synchronous speed, a frequency-sensitive control

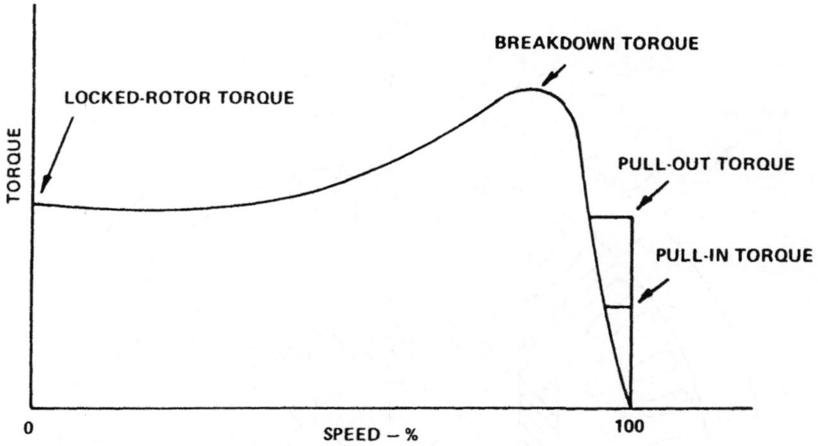


**Figure 6.11.** Speed–torque curve of a synchronous motor.

disconnects the discharge resistor and applies excitation at the proper speed and phase angle to obtain maximum pull-in torque. It is desirable not to allow the motor to "slip" poles during the pull-in process as this can cause large transients in the power supply line and excessive motor vibration.

From an energy conservation standpoint, synchronous motors should be considered in the large horsepower sizes or for slow-speed applications where the combination of high efficiency and high power factor can justify the increased complexity of motor and control.

## 6.4. Permanent-Magnet Synchronous Motors

Permanent-magnet (PM) motors have grown in both numbers and importance ever since the introduction of ceramic and high-energy-density magnets. For many years PM motors utilized alnico magnets which were vulnerable to demagnetization under overload and fault conditions. The availability of ceramic, samarium cobalt, and, more recently, neodymium iron magnets has improved the performance of PM motors and they are now being utilized in a much wider range of applications.

Figure 6.12 is a schematic view of one type of PM synchronous motor showing the essential construction features. The stator is very similar to a conventional induction motor stator with a polyphase

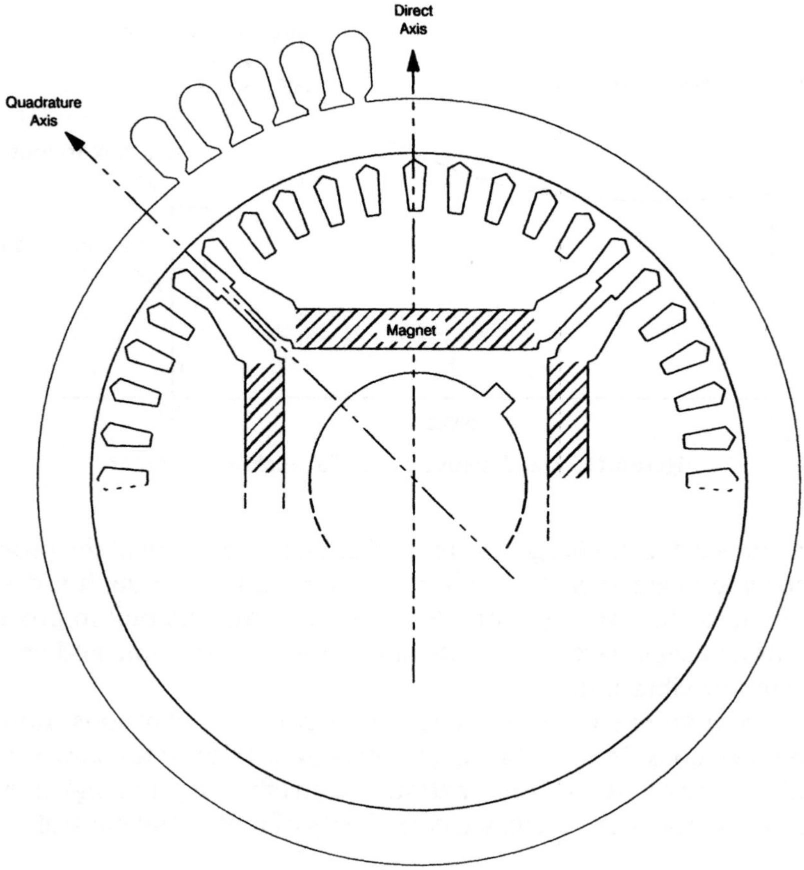


**Figure 6.12.** Permanent-magnet synchronous motor using buried magnet construction (U.S. Patent No. 4,139,790; Charles R. Steen assigned to Reliance Electric Co.).

winding designed to establish a rotating magnetic field when excited with a balanced set of three phase voltages. The rotor, shown in Fig. 6.12, contains four magnetic poles surrounded by a squirrel-cage winding which is short-circuited at both ends, similar to a conventional induction motor. The motor accelerates from rest as an induction machine, which is the reason for including the squirrel cage in the rotor construction. As the rotor rotational speed nears synchronous speed, the rotor pulls into synchronism and operates thereafter as a synchronous motor. During steady-state operation, there is no current in the rotor, thus no losses, and this contributes to the high efficiency of the PM synchronous motor.

This type of PM synchronous motor is sometimes referred to as a buried magnet construction because the magnets are buried beneath the rotor surface. The figure shows the rotor with the magnet secured in place by slots in the lamination. The slots extending from the magnet cavity toward the periphery of the rotor aid in defining the direct axis magnetic path as a distinctly different magnetic path from the quadrature axis. The torque for the machine is thereby increased because of the saliency torque which is proportional to the difference between the reactances in the two axes. In this construction, the magnet is surrounded by die-cast aluminum which fills the rotor slots.

A second, commonly used construction is shown in Fig. 6.13. The magnets are secured to the rotor surface and, in this design, there is no

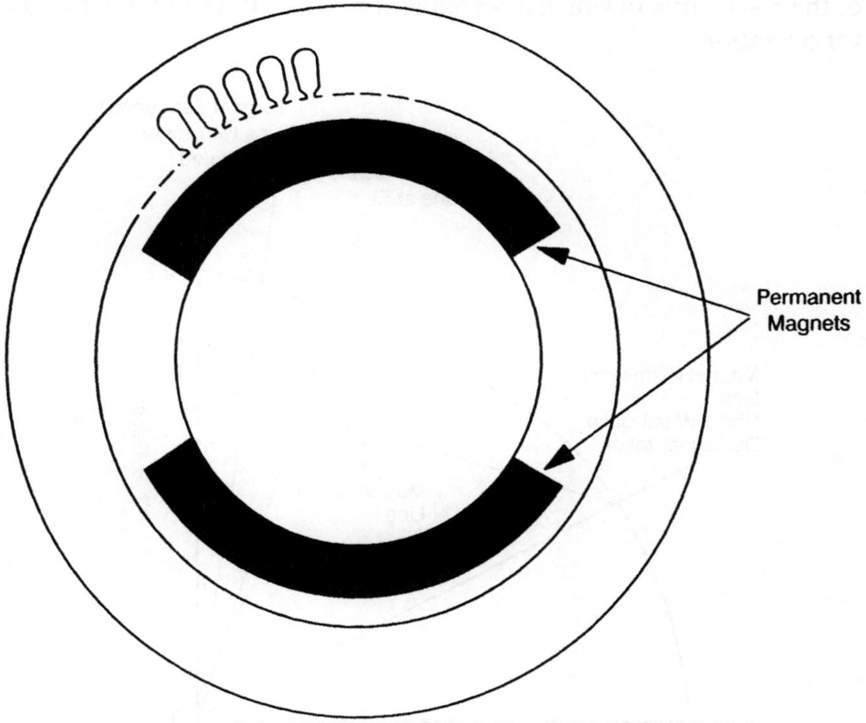


**Figure 6.13.** Permanent-magnet synchronous motor using surface mounted magnets.

squirrel cage to provide induction motor-type acceleration. Therefore, the application of this design usually requires an adjustable-frequency power supply. The motor is locked into synchronism at a low frequency and then proceeds to its final operating speed by increasing the frequency. The absence of the squirrel cage means that the motor must remain operating in synchronism at all times. If the rotor pulls "out of step" as a result of an overload or other condition, there is no recovery other than to stop the motor and resynchronize it.

This type of motor construction is widely used in small brushless dc fan motors and also in brushless dc servo motors. More details of the control system associated with brushless dc motors are given in Section 7.8.

The very strong influence that various types of permanent magnet materials have on motor designs is best understood by considering the *B–H* curves of a permanent magnet. Figure 6.14 shows the region of the *B–H* curve in which a permanent magnet functions during motor operation.

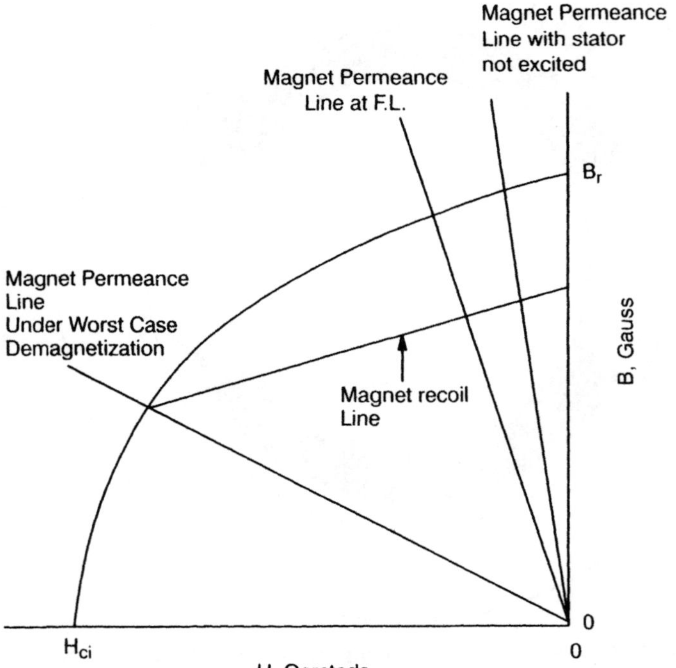


**Figure 6.14.** Magnet operating lines on a permanent-magnet *B–H* curve.

Three permeance lines corresponding to three different conditions of operation are shown:

1. Stator not excited
2. Full load
3. Maximum demagnetization—depending on the motor design, this could be pull-in, pull-out, starting, etc.

Also shown in Fig. 6.14 is the "recoil line." Once the magnet has been demagnetized to a particular point on the B–H curve, the magnet recovers along a recoil line which approximates the slope of the B–H curve in the region near $H = 0$. Since operation is at the intersection of a permeance line and the *B–H* curve, or recoil line, it is evident that the point of greatest demagnetization is critical to the motor's performance.

Now consider the typical *B–H* curves for the four different types of permanent magnets shown in Fig. 6.15. These four materials are:

- Alnico
- Ceramic
- Samarium cobalt
- Neodymium iron

The preceding discussion about demagnetization and the "recoil line" provides an understanding as to why alnico is a difficult material to utilize in motor design. Any demagnetizing force which causes the effective magnet permeance line to drop below the knee of the demagnetization curve will cause the recoil line to return to a residual magnetization (flux density at $H = 0$) point much lower than its initial value. Thus, the torque of the motor changes with operation until it stabilizes at some value corresponding to the maximum demagnetizing force.

The other three types of magnets, although not totally immune to this effect, are not as dramatically affected by demagnetization.

A parameter which is useful for characterizing permanent magnets is the energy product. The energy product at any point on the demagnetization curve is the product $B \times H$. Clearly, the energy product is zero at points where the curve crosses the $B$ and $H$ axes. These two points are designated as $H_c$, the coercive force, and $B_r$, the residual magnetization. In between, the energy product reaches some maximum value $(BH)_{max}$. The maximum energy product is frequently

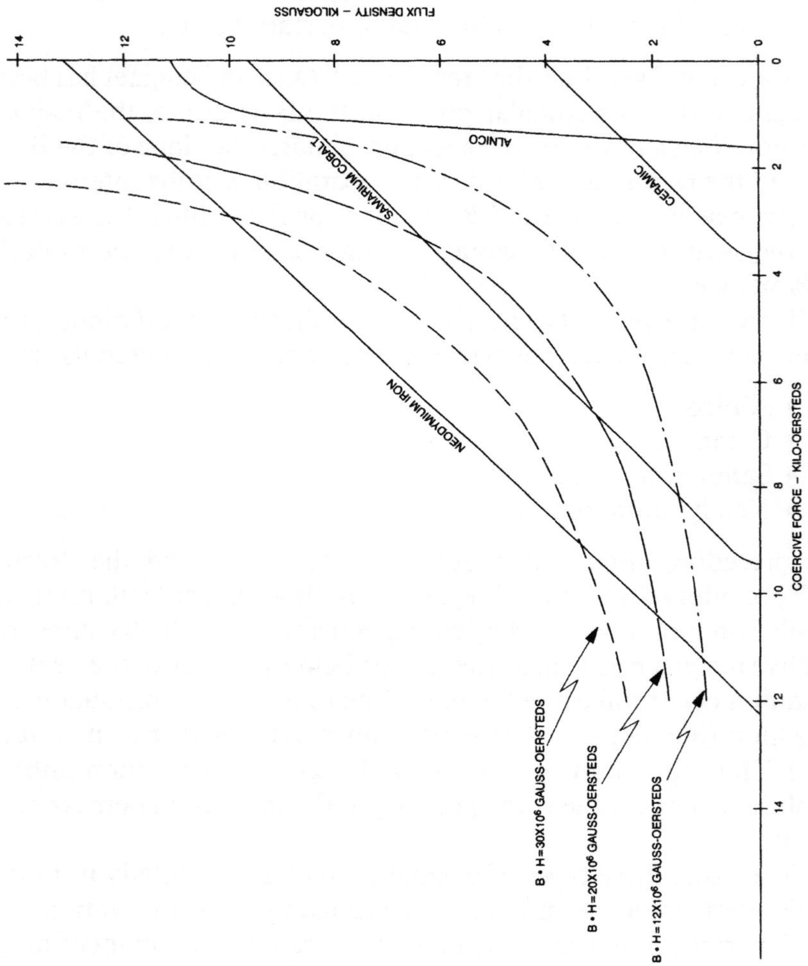


**Figure 6.15.** Permanent-magnet demagnetization curves.

used as a means of characterizing and comparing different magnets. The higher this product, the stronger the magnet is. For the most effective utilization of magnet material, operation should be at the point where $(BH)_{max}$ occurs. In motor designs it is frequently not possible to operate right at the $(BH)_{max}$ point, at least not for all loads. Nevertheless, the $(BH)_{max}$ parameter is useful in comparing and selecting the type of magnet material to be used.

Frequently the *B–H* curves published by permanent magnet manufacturers will display a series of hyperbolic curves labeled as energy product curves superimposed on the magnet demagnetization curves. Three such curves are shown in Fig. 6.15. The energy product for any point on the demagnetization curve is given by its intersection with the energy product curve at that point. The $(BH)_{max}$ value is given by the energy product curve which is tangent to the demagnetization curve. Since an exact tangent usually does not occur, given the discrete number of energy product curves plotted, some interpolation is usually required. In Fig. 6.15, the maximum energy product for the samarium cobalt curve is approximately $24 \times 10^6$ gauss oersteds.

Table 6.1 lists the three permanent magnet types commonly used in motors and their maximum energy products plus a temperature coefficient number.

Samarium cobalt and neodymium iron are both high-energy materials compared with ceramic magnets. The temperature coefficient of the residual induction is included in the table because it focuses attention on the fact that the magnet performance at operating temperature can be quite different from room temperature performance. Neodymium iron has a substantially larger temperature coefficient than samarium cobalt. The temperature effect on the coercive force must also be considered. Actually the entire demagnetization curve at operating temperature must be known to evaluate a particular magnet for a motor design. A complete discussion of temperature effects is beyond the scope of this chapter but suffice it to say that magnet

**Table 6.1. Properties of Permanent-Magnet Materials**

| Property | Samarium cobalt | Neodymium iron | Ceramic |
|---|---|---|---|
| $(BH)_{max}$ (gauss oersted) | $15–28 \times 10^6$ | $30–40 \times 10^6$ | $3–5 \times 10^6$ |
| Temperature coefficient of residual induction (%/°C) | −0.04 | −0.12 | −0.20 |

performance can be strongly influenced by the internal operating temperature of the motor.

The other dimension of motor magnet selection is the cost of the magnet. Without considering cost, one would unhesitatingly select one of the high-energy magnets, samarium cobalt or neodymium iron. However, the cost per pound of a high-energy magnet can be between 10 and 100 times the cost of a ceramic magnet. Consequently, the high-energy magnets are usually used in applications where performance and size are critical. Servomotors and aircraft motors are good examples. Cost-sensitive applications such as fan motors frequently utilize ceramic magnets.

Permanent magnet synchronous motors offer the potential of being more energy efficient than induction motors because the excitation energy does not have to be supplied from the power lines. Richter *et al.*[6] have studied this matter including the economics of offsetting the higher first cost of the motor with energy savings during the operating lifetime of the motor.

Chapter 2 discusses the five components of loss in an induction motor. The PM motor can reduce two of these, namely,

- Primary $I^2r$
- Secondary $I^2r$

The primary $I^2r$ loss reduction occurs because the excitation is provided by a permanent magnet rather than by a magnetizing current in the stator winding. The secondary $I^2r$ loss is eliminated because the motor operates in synchronism and consequently no slip power is dissipated in the rotor. Offsetting this, Schiferl *et al.*[7] have observed that the core loss in a PM motor can be larger than in a comparable induction motor. Although efficiency data on PM synchronous motors are somewhat sparse, loss reductions of 15–30% relative to the equivalent energy-efficient induction motors are typical. This could arguably offset the added cost of materials and labor to construct the PM motor compared with its induction motor counterpart. However, in reality, the first cost penalty for purchasing a PM motor has been sufficiently large that these motors have found very little application purely as energy-saving replacements for induction motors. PM motors are, however, widely used in brushless dc drives (see Chapter 7).

One final characteristic worth mentioning is the PM synchronous motor's pulsating torque during starting. This consideration is important for across-the-line starting which is the case for the vast majority of motor applications. Since the permanent magnet field is always present, it produces an oscillatory torque during the acceleration period up to the time that the rotor is synchronized. Richter *et al.*[6] report peak transient torques of 600% of rated torque for a particular 25-hp, 1800-rpm design. Induction motors also exhibit transient torques during start-up. However, the PM synchronous motor transients are somewhat more severe and they last all the way to synchronous speed whereas induction motor transients disappear at about half speed.

A common application area for PM synchronous motors is in multimotor, variable-speed drives where precise synchronization between driven sections is important. The synthetic fiber industry uses large numbers of small synchronous motors to drive the multiple spindles which draw synthetic fibers and wind them onto spools. The glass industry also requires synchronized drives which use this type of motor.[4]

In variable-speed applications, efficiency and power factor are important for two reasons, namely,

1. Energy costs are reduced by improved efficiency.
2. Installed cost of power converter is proportional to motor kVA.

Variable-speed ac drives are supplied from solid-state power converters, usually inverters which provide the variable frequency and variable voltage output necessary to achieve variable speed. The size and cost of these power converters are determined by the kVA that they must supply. Therefore, an improved-performance motor has a double economic impact. Not only is energy saved because of reduced motor losses, but the capital cost of the power unit is reduced because of the reduced kVA input requirements of the motor.

## References

1. Veinott, C.G. *Theory and Design of Small Induction Motors.* New York: McGraw–Hill Book Co., Inc., 1959.
2. Veinott, C.G. *Fractional Horsepower Electric Motors, 2nd ed.* New York: McGraw–Hill Book Co., Inc., 1948.

3. Fitzgerald, A.E., and Kingsley, C. *Electric Machinery.* New York: McGraw–Hill Book Co., Inc., 1952.
4. Jordan, H.E. Permanent magnet synchronous motors in glass industry applications, *IEEE, IAS Annual Conference,* 1977.
5. Honsinger, V.B. U.S. Patent No. 3,045,135, "Synchronous Induction Motor."
6. Richter, E., Miller, T.J.E., Neumann, T.W., and Hudson, T.L. The ferrite permanent magnet AC motor—A technical and economical assessment, *Conference Record, 1984 IEEE IAS Annual Meeting.*
7. Schiferl, R.F., Colby, R.S., and Novotny, D.W. Energy considerations in permanent magnet synchronous motor drives, *Electrical Energy Conference,* Adelaide, Australia, October, 1987.

# 7

# Adjustable-Speed Drives

## 7.1. Adjustable Speed Applied to Pumps and Fans

Adjustable-speed drives have been a stalwart contributor to the success of many of the world's industrial processes for years. Textile fiber manufacturing, steel mill rolling, automobile tire cord processing, and a myriad of other processes utilize adjustable-speed electric drives in the manufacturing process. High energy costs are now causing interest to be focused on adjustable-speed drives for pumps and fans.

Figure 7.1 displays a typical pump characteristic curve of flow versus pressure. On the same graph, the curve for the system into which the pump is delivering its fluid is also shown. The system characteristic is comprised of two parts: a static head or lift which the pump must supply, and the friction loss as the fluid passes through the pipes. This friction loss obeys a square law,

$$\text{Pressure} = K\,(\text{flow})^2 \qquad (7.1)$$

A widely used method for controlling flow is to throttle the flow somewhere along the piping system. The effect of this type of control can be shown graphically as in Fig. 7.1(a) which displays two flow conditions. Point A is the intersection of the unthrottled system curve with the pump curve and this represents the maximum delivery which can be achieved for this particular combination of pump static head and friction flow resistance. When a reduction in flow is desired, the flow is throttled so that it operates at Point B. In Fig. 7.1(b) the power

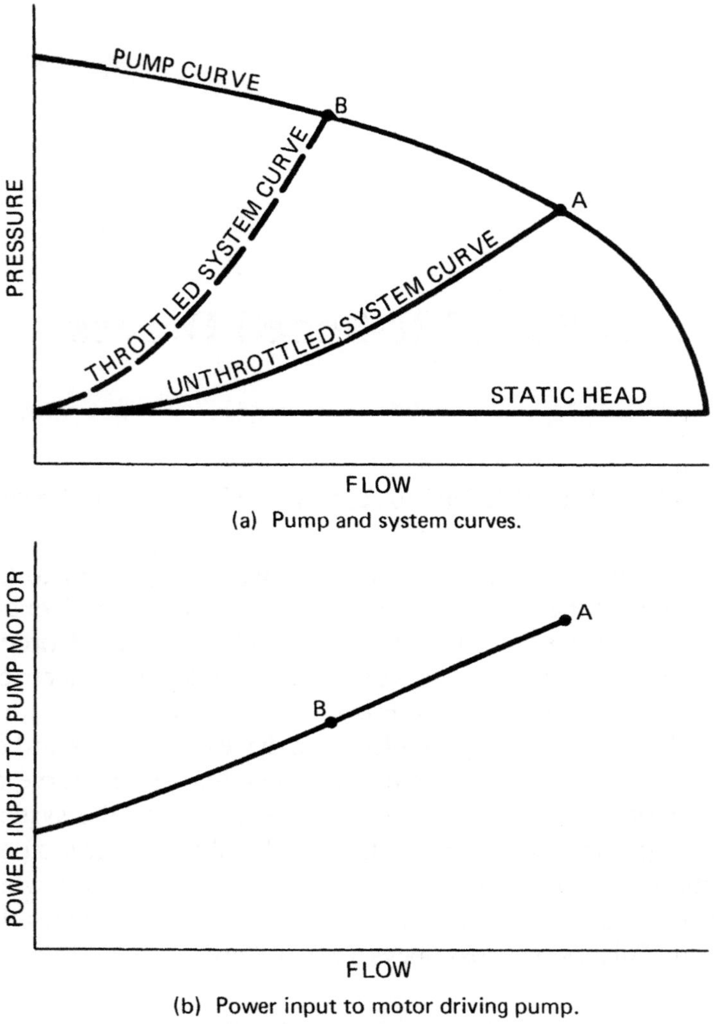


**Figure 7.1.** Characteristic curves for a pump driven by a constant-speed motor.

input to the motor driving the pump is shown and the inputs associated with conditions A and B are indicated.

The situation pictured in Fig. 7.1 applies to a pump driven by a constant-speed ac motor, and a similar analysis applies to a fan when it is throttled on its output side, since both pumps and fans obey the square law of Eq. (7.1). In the case of a fan, there is no static head to be supplied and the system curve originates at the origin. In some ventilating systems, air flow is regulated by adjusting vanes or dampers on the inlet side of the fan. In these cases, the energy savings achievable by speed control are not quite as dramatic as those from outlet throttling, but savings potential nevertheless exists in both cases.

The term "constant" speed is slightly in error, because if the pump is driven by an induction motor, the slip, of course, does vary with load but the percentage change in speed is small and can be neglected.

An alternate, and more energy-efficient means of accomplishing flow control is to vary the speed of the pump. The flow and pressure vary in proportion to the speed and $(\text{speed})^2$, respectively:

$$\text{Flow} = \text{A} \times (\text{pump speed}) \tag{7.2}$$

$$\text{Pressure} = \text{B} \times (\text{pump speed})^2 \tag{7.3}$$

Figure 7.2(a) shows how a speed change can be used to adjust the flow and Fig. 7.2(b) shows the power input into the adjustable-speed drive unit used to provide the variable speed. Figure 7.3 displays the power input curves of Figs. 7.1(b) and 7.2(b) on a single graph, and the dashed curve is the difference between the two power input curves. This dashed curve represents the power savings achievable by using adjustable speed rather than throttling for flow control.

In essence, throttling wastes energy as it dissipates energy which has been put into the fluid by the pump in the throttling valve. An adjustable-speed drive simply reduces the total energy into the system when it is not needed.

The purchase and installation of a throttling valve is usually less costly on a first-cost basis than the first cost of an adjustable-speed drive. So, as in most energy conservation situations, trade-offs between first cost and annual energy savings must be evaluated and this has to be done on an individual application basis. Applications requiring extended operating periods at reduced flow rates will be most favorable for selection of an adjustable-speed drive.

Fans and pumps do not present the only opportunities for energy savings by variable speed. Processes which require braking effort and where the braking energy is dissipated in the form of heat are also candidates. Test stands often fall into this category. An example of a

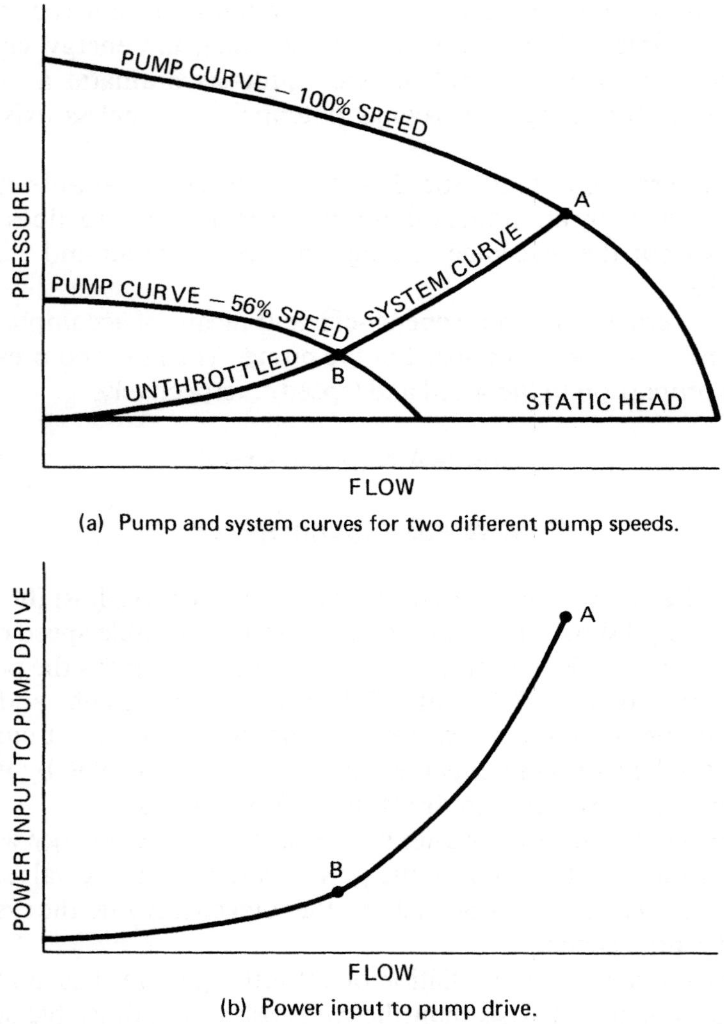


**Figure 7.2.** Characteristic curves for a pump driven by an adjustable-speed drive.

testing application is the dynamometers which are widely used for testing engines and transmissions. By use of an adjustable-speed inverter drive, the energy from the dynamometer, which is serving as a brake for the unit being tested, can be returned to the power line and the energy cost for testing is reduced to only the cost of supplying the system losses. Process lines requiring that tension be maintained on material being processed can save energy by replacing mechanical brakes with regenerative, adjustable-speed drives.

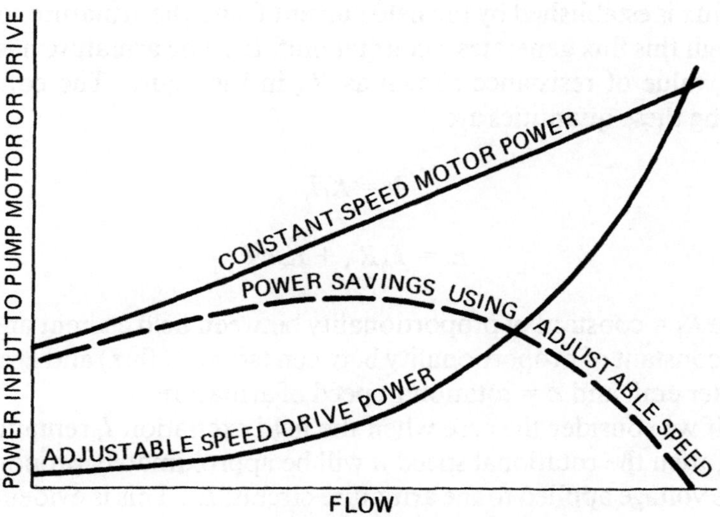


**Figure 7.3.** Power savings with adjustable-speed pump drive.

Because of the potentials for energy savings that exist with variable speed, a discussion of several of the most common methods of achieving variable speed is presented in the following sections.

## 7.2. dc Drives

Direct current drives date back to the 1930s when the Ward-Leonard system was patented by H. Ward-Leonard. They have undergone many evolutionary changes over the years, but today the dc drive remains one of the most versatile and widely used methods of speed control. Figure 7.4 shows the equivalent circuit of a dc motor.

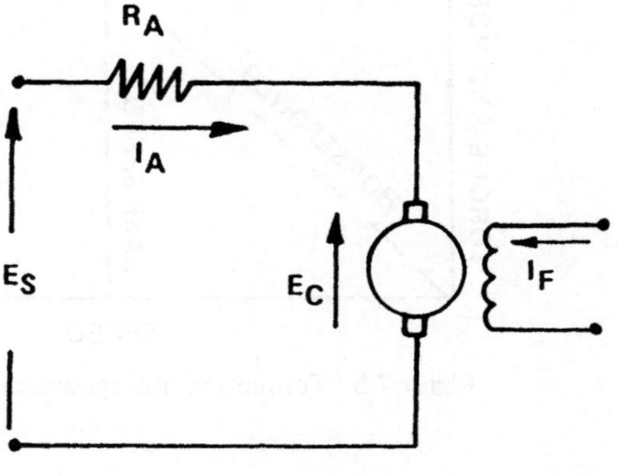


**Figure 7.4.** Equivalent circuit of dc motor.

The flux is established by the field current $I_f$ and the armature rotating through this flux generates a counter emf, $E_c$. The armature has some finite value of resistance shown as $R_A$ in the figure. The equations relating these quantities are

$$\Phi = K_f I_f \tag{7.4}$$

$$E_s = I_A R_A + K_e n \Phi \tag{7.5}$$

where $K_f$ = constant of proportionality between field current and flux, $K_e$ = constant of proportionality between (speed × flux) and armature counter emf, and $n$ = rotational speed of armature.

If we consider the case when the field excitation $I_f$ remains constant, then the rotational speed $n$ will be approximately proportional to the voltage applied to the armature circuit, $E_s$. This is evident from Eq. (7.5) if one recognizes that the term $I_A R_A$ is small with respect to the term $K_e n \Phi$ in the equation. This mode of control is referred to as armature voltage control and is the mode used from the lower end of the speed range up to base speed as shown in Fig. 7.5. In this region, the motor can deliver approximately constant torque, and, therefore, the horsepower increases directly with speed. Above base speed the motor operates in a constant-horsepower region. The armature volt-

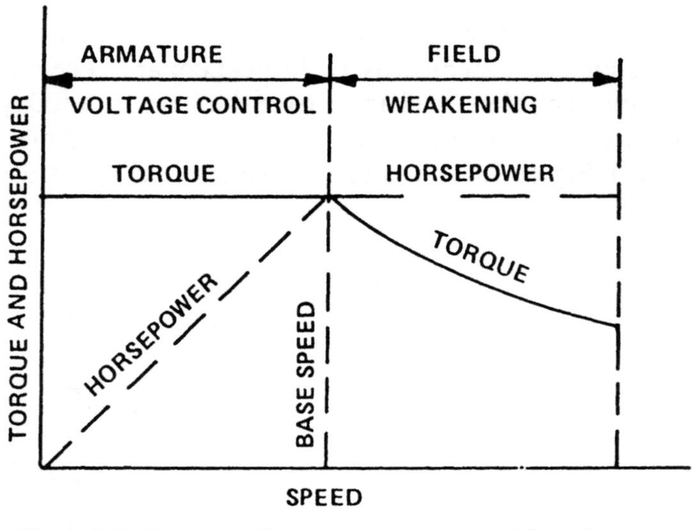


**Figure 7.5.** Torque and horsepower versus speed for a dc motor.

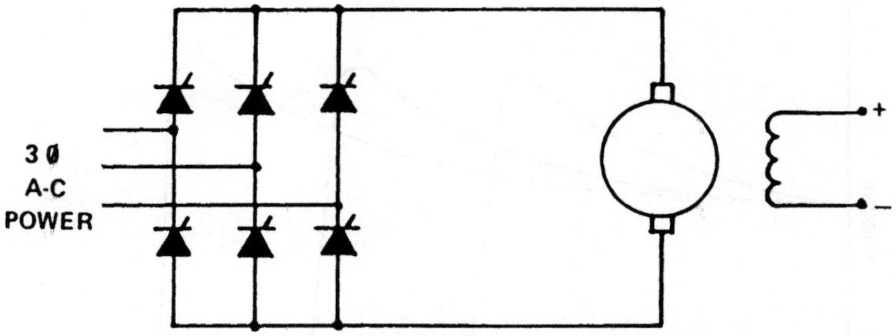


**Figure 7.6.** Controlled rectifier and dc motor.

age is maintained constant, and the field is weakened to cause the motor speed to increase. However, with the weakened field, the torque carrying capability of the motor declines and, therefore, it is a constant-horsepower drive above base speed.

In both pump and fan applications, the armature voltage range is the applicable range of operation. Since most pump and fan applications require only a 2:1 or a 3:1 speed range, dc drives can readily meet these application requirements. Many dc drives are capable of a considerably wider speed range, 6:1 is typical, and are also capable of delivering more torque than a fan or pump needs at low speed. For this reason, a high-performance dc drive is not required for a pump or fan application, and the economic payback will be more favorable if the drive is specified to provide only the performance characteristics needed by the fan or pump.

Figure 7.6 shows one of the commonly used configurations for a dc motor drive. A controlled rectifier bridge operating directly from a three-phase ac power line provides a dc voltage for the motor armature. By controlling the phase angle at which the thyristors in the bridge are placed in conduction, the effective voltage on the motor's armature is controlled and thereby variable speed is achieved.

## 7.3. ac Drives—Inverters

Solid-state inverters are the most common means of achieving variable-speed ac operation. Figure 7.7 displays a family of induction motor speed–torque curves as would be obtained by powering an in-

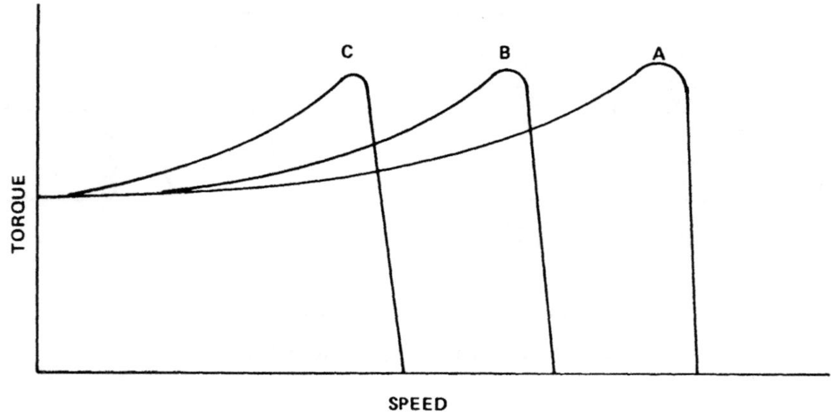


**Figure 7.7.** Speed–torque curves for induction motor operating from variable frequency and variable voltage source.

duction motor from a variable-frequency and variable-voltage source. Figure 7.7 shows three specific speed–torque curves, but most inverter drives provide infinitely variable speed within the operating speed range. Both voltage and frequency must be varied in some defined relationship to one another if the motor is to provide a constant torque capability over the speed range, and most inverter drives provide this capability over at least a portion of their speed range. Some drives provide a "constant" horsepower range at the upper end of the speed band, and in these cases, the frequency is increased while the voltage is maintained constant. The base speed at which this transition from constant torque to constant horsepower takes place is usually 60 Hz.

The circuit diagram for one type of inverter is shown in Fig. 7.8. The circuit is composed of three distinguishable elements, namely:

- Controlled rectifier
- dc link
- Inverter

The controlled rectifier rectifies the incoming three-phase ac power to dc, and, by varying the phase angle for firing of the thyristors, controls the dc voltage level. The dc link filters the ripple at the output of the rectifier and the combination of the controlled rectifier and filter provides a variable dc voltage to the inverter. The inverter then provides three-phase ac power to the motor by proper gating of the

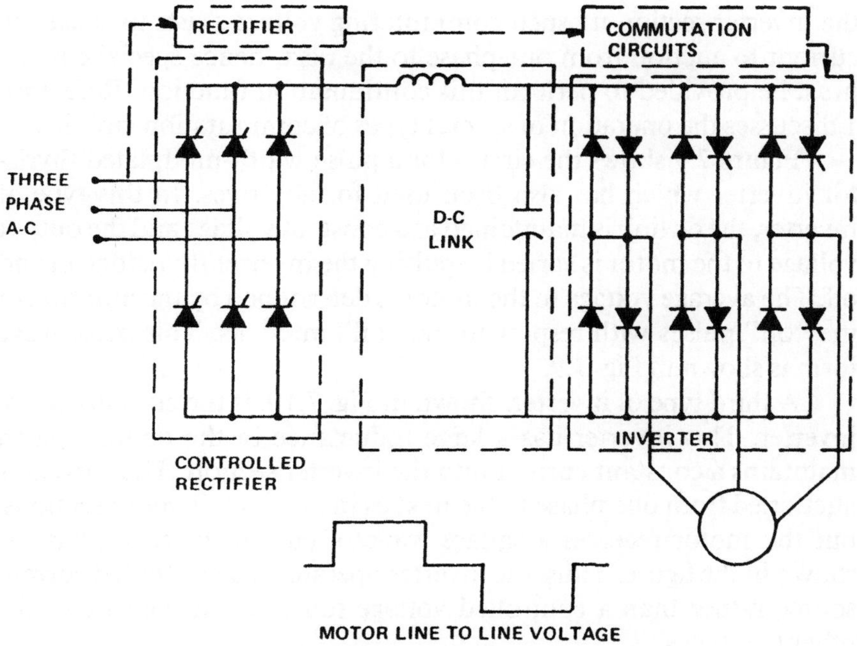


**Figure 7.8.** Variable-voltage input inverter.

thyristors in the inverter unit. More detailed descriptions of inverter action and of the various types of inverters described in this section are given in Reference 1.

One of the features which distinguishes between the various inverter types is the means of controlling the voltage so that it maintains some fixed relationship with the frequency. It is usual to vary the voltage in direct proportion to the frequency except at the very low end of the frequency range where some offset is provided to improve the starting torque. The circuit shown in Fig. 7.8 is called a variable-voltage input inverter because the voltage control is achieved at the dc link. The output of the controlled rectifier is varied with the frequency so that the motor voltage and frequency remain in the desired relationship.

A thyristor once turned on will not go out of conduction so long as a current greater than the holding current is present in the forward direction. In the controlled rectifier section, the alternating half cycles of the three-phase ac line provide the reverse bias across each thyristor to turn it off and auction the current to the next phase. However, in the inverter section no such commutating voltage exists to cause the current to auction from one phase to the next, hence special circuits must be provided to perform this commutation function. Reference 1 discusses the operation of several types of commutation circuits.

Figure 7.9 shows the circuit for a pulse-width-modulated thyristor inverter which has also been used for ac drives. In this type of inverter, the dc link is maintained at a constant voltage and the output voltage to the motor is varied by pulsing the inverter thyristors on and off. The average voltage to the motor is determined by the duration of the "on" pulses with respect to the "off" intervals. This pulse wave form is shown in Fig. 7.9.

A third type of inverter, shown in Fig. 7.10, is the current-source inverter. This inverter has a large inductance in the dc link which maintains a constant current into the inverter section. The current is auctioned from one phase to the next as in the other inverter sections, but the motor receives a square wave of current in each phase as shown in the figure. Thus, the inverter operates as a controlled current source rather than a controlled voltage source as in the case of the other two types.

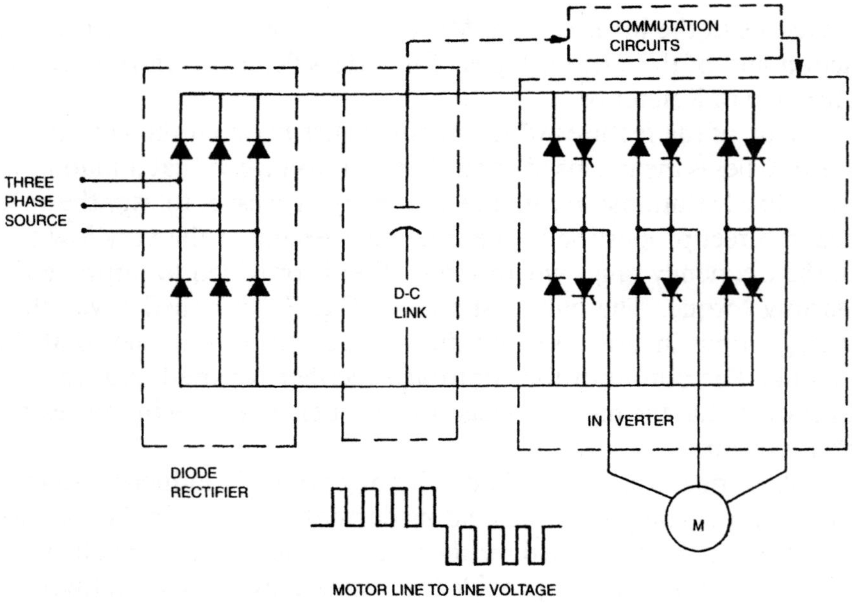


**Figure 7.9.** Pulse-width-modulated thyristor inverter.

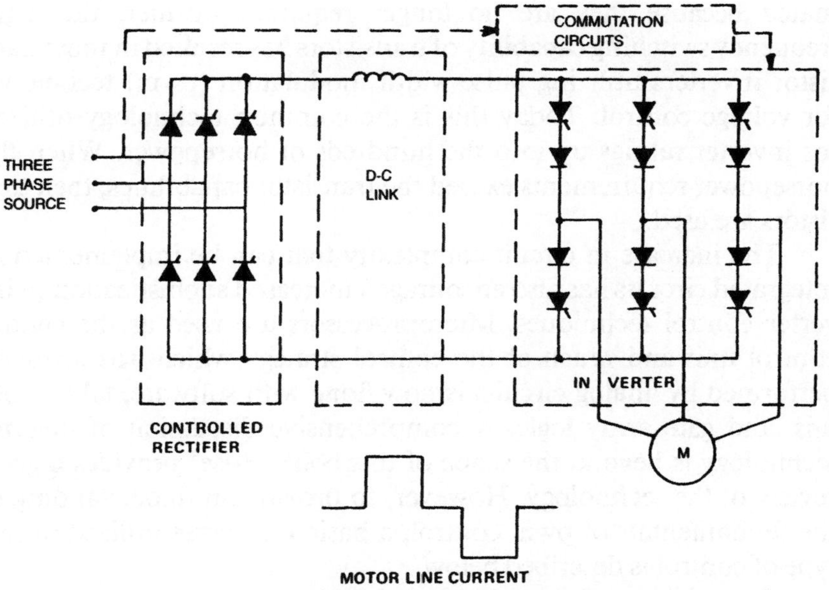


**Figure 7.10.** Current-source inverter.

Each of the three inverter types has its own particular advantages and disadvantages. All are in use on adjustable-speed applications and an individual evaluation is necessary to determine which type should be selected for a particular requirement.

## 7.4. Transistor Pulse-Width-Modulated Inverters

Inverter technology has changed over time influenced by three factors:

- Current and voltage ratings of solid-state power switches
- Integrated circuit implementation of inverter control circuits
- Application of high-speed microprocessors to inverter control

The foregoing sections described the basic types of inverters utilizing thyristor power switches in the inverter power section. As transistors have increased in their voltage and current ratings, they have replaced thyristors as the power switches of choice in many inverters. Along with this transition, the forced commutation circuits have been eliminated because they are no longer required. Further, the high-frequency switching capability of transistors has resulted in most transistor inverters utilizing pulse-width modulation (pwm) techniques for voltage control. Today this is the common technology utilized for inverter ratings up into the hundreds of horsepower. When the horsepower requirements exceed the transistor capabilities, then thyristors are used.

The increase in circuit complexity that can be implemented in integrated circuits has also encouraged increased sophistication in inverter control techniques. Microprocessors are used as the central control unit and much of the control strategy which was formerly performed by analog circuits is now done with software, table look-ups, and gate array logic. A comprehensive discussion of inverter technology is beyond the scope of this book. Bose[4] provides a good review of the technology. However, to provide an understanding of the fundamentals of pwm control, a basic technique utilized in this type of control is described below.

One of the earliest techniques for generating a pwm waveform which is still used today is the "sine-triangle" technique described by Mokrytzki[5] and many others. Variations of this basic technique are in use but the sine-triangle modulation technique provides an understanding of the fundamentals of pulse-width modulation.

Figure 7.11 displays a transistor inverter circuit operating from a constant-voltage dc bus. For the purposes of explaining the inverter circuit operation, the dc bus is shown as two batteries and a center tap. In actual inverters the dc bus is usually established by a three-phase, full-wave rectifier and the center tap is really not required. The sine-triangle technique for generating the proper pulse configuration is illustrated in Fig. 7.12. The triangle carrier waveform with two phase reference voltages are shown in Fig. 7.12(a). The third phase is omitted to simplify the drawing.

Consider the voltage $V_{ao}$ in Fig. 7.12(b). Switching between the upper transistor in phase A, $Q_1$, and the lower transistor, $Q_4$, occurs each time $V_a$ crosses the carrier wave. This switching sequence is shown in Fig. 7.12(b). Similarly, $V_{bo}$ is governed by the crossings of $V_b$ and the carrier as shown in Fig. 7.12(c). $V_{ab}$ is obtained by:

$$V_{ab} = V_{ao} - V_{bo} \tag{7.6}$$

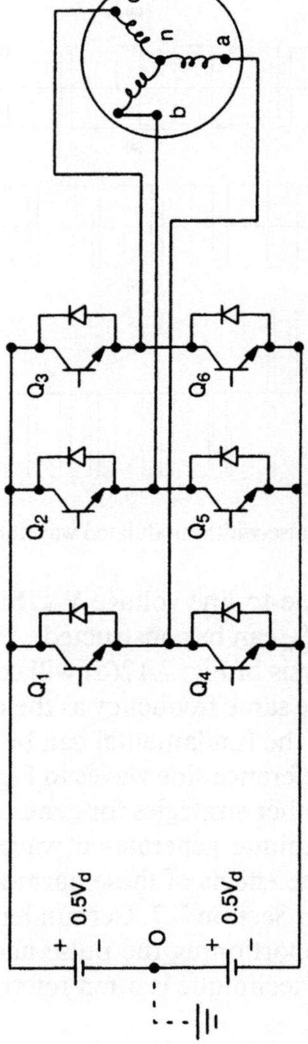


**Figure 7.11.** Three-phase transistor inverter.

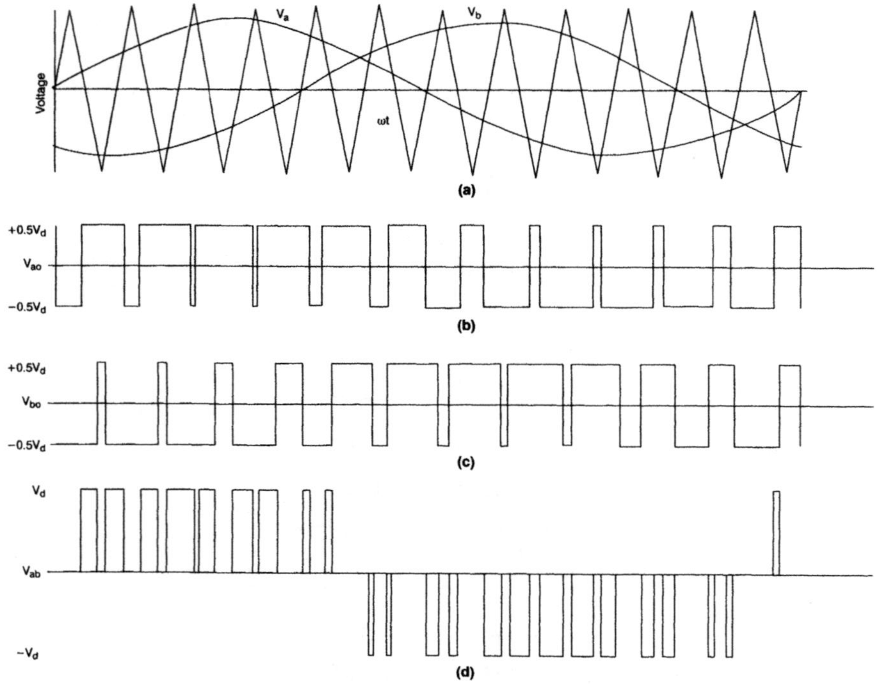


**Figure 7.12.** Pulse-width-modulated waveform generation.

Figure 7.12(d) is the line-to-line voltage $V_{ab}$. Similarly, the other two line voltages $V_{bc}$ and $V_{ca}$ can be constructed.

A harmonic analysis of Fig. 7.12(d) will confirm that the fundamental frequency is the same frequency as the reference voltage. Further, the amplitude of the fundamental can be controlled by varying the amplitude of the reference sine waves in Fig. 7.12(a).

There are many other strategies for generating pwm waveforms. The sine-triangle technique generates a waveform with numerous harmonics present. The effects of these harmonics on motor performance are discussed in Section 7.7. Certain harmonics can be eliminated by properly proportioning the pulse and notch widths as described by Bose.[4] This technique is often referred to as the harmonic elimination method.

Microprocessors are also used to generate the pwm modulating signals. Some algorithms implement the sine-triangle technique just described by repetitively solving for the intersection of the sine wave reference and the triangle carrier. This technique is computation intensive and limits the switching frequency that can be employed because of the time required to calculate each switching event. A more frequently utilized algorithm stores the pulse and notch widths in a lookup table. The table data are loaded into down counters and the transistor on/off signals are output when the counters count down to zero. The clock frequency for the counter is controlled by the desired fundamental frequency.

Despite the harmonics inherent in pwm waveform generation, pwm inverters generally provide very acceptable output waveforms for operating induction motors and are, therefore, widely used. The inverters used for pump and fan applications can be relatively simple, from a control hierarchy point of view. There are many other applications for inverters such as machine tool drives, positioning systems, etc., which require a more sophisticated control. This class of applications demands rapid response to a change in a speed or position reference signal. The inherent characteristics of an inverter-induction motor drive are not favorable for high-response applications. However, by adding supplemental control paradigms to a basic inverter, a highly responsive inverter-induction motor drive can be built and, indeed, many such drives are available commercially.

One of the widely used techniques to achieve the desired responsiveness is called vector or field-oriented control. It addresses the basic problem of trying to adjust the speed of an induction motor too rapidly.

Consider a motor operating at point P1 on curve A in Fig. 7.13. Now, if the frequency suddenly increases to change the speed–torque curve to some new curve B and the motor speed, because of a high load inertia, cannot increase rapidly, the motor will operate at some new point P2. However, P2 is at a high slip relative to curve B. The motor will demand more current and, depending on the inverter control, may stall or cause the inverter to trip off of the line because of the high current demanded by the inverter.

Vector control avoids this scenario by reconfiguring the control so that the flux and armature mmf fields in the motor do not attain this high-slip condition. The underlying idea behind a vector control system is to decouple the control of flux and motor torque and to control the two independently as one does in a dc motor. The mathematical basis for this and the control schemes to implement this type of control are described by Boldea and Nasar,[6] Vas,[7] and many others.

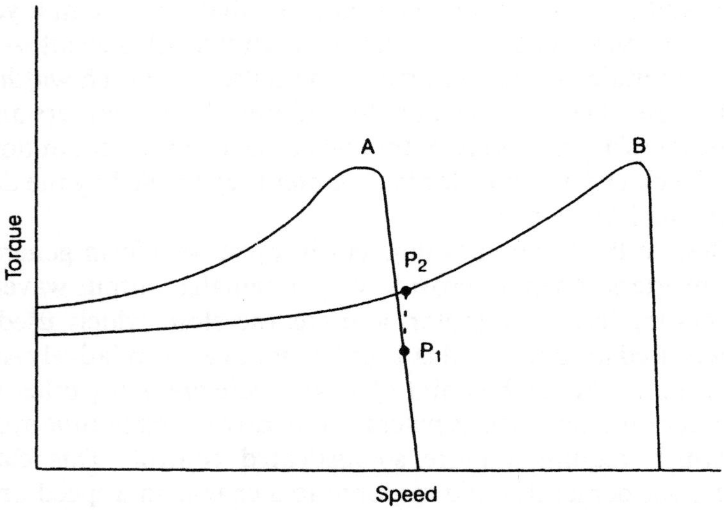


**Figure 7.13.** Induction motor speed–torque curves for two different operating frequencies.

The control complexity is certainly increased relative to the more basic volts per hertz control but the system responsiveness is also much improved and this is important for many applications.

## 7.5. Load-Commutated Inverter Drive

Among the family of ac motor drives, the load-commutated inverter drive possesses some characteristics which make it unique from the inverter-induction motor drives discussed in the preceding section. The circuit diagram for this drive is displayed in Fig. 7.14, and it is notable because of the lack of commutation circuits. This type of drive utilizes natural rather than forced commutation because the emf generated at the armature terminals is available to commutate the thyristors. Of course, an exciting field is required to provide this emf, so a synchronous motor is used in place of an induction motor. This somewhat added complexity for the motor permits a simplification in the electronics because forced commutation circuits are not required.

One version of this type of drive derives the inverter frequency signal from a shaft-mounted speed sensor so that the drive is always

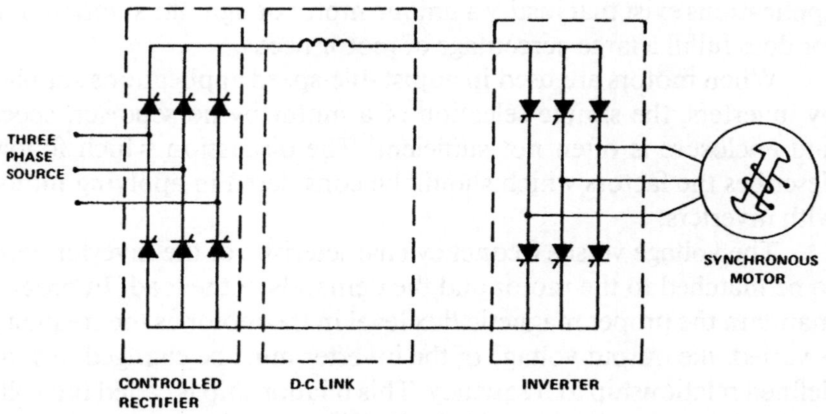


**Figure 7.14.** Load-commutated inverter drive.

operating in synchronism. A command to increase or decrease speed causes the dc link voltage to be increased or decreased, and thus this drive has been likened to a dc drive which has armature voltage as its primary mode of control.

The elimination of forced commutation is particularly advantageous for large machines designed to operate at 2300 volts and higher, so the load-commutated inverter drive is particularly attractive in large horsepower sizes.

## 7.6. Frequency–Voltage Relationships and Starting for Motors Used with Adjustable-Speed Drives

The engineering required to successfully apply a fixed-frequency, 60 Hz, motor to the driven equipment may, in many cases, be simply the selection of the correct horsepower, speed, and motor enclosure. The ease with which this task can be performed is related to the fact that motors are available from many suppliers which meet NEMA.[8] Standards for torques, inrush current, temperature rise, etc. Mindful that motors which meet these standardized specifications are readily available, equipment manufacturers design their products to be compatible with standard motors, thus minimizing the users' task in selecting and purchasing a motor. While many definite purpose motor applications exist that justify a unique motor design, the standard motor does fulfill a large percentage of motor needs.

When motors are used in adjustable-speed applications supplied by inverters, the simple selection of a motor by horsepower, speed, and enclosure is often not sufficient. The discussion which follows describes the factors which should be considered in applying motors with inverters.

The voltage versus frequency characteristic of the inverter needs to be matched to the motor and the demands of the load. In order to maintain the proper magnetic flux level in the motor as the frequency is varied, the output voltage of the inverter must be changed in some defined relationship to frequency. This relationship is called the volts-per-hertz curve which is established by the control logic of the inverter and is often adjustable so that it can be tailored to the application. A method for determining the optimum voltage and frequency relationship has been described by Tsivitse and Klingshirn.[9]

A typical volts-per-hertz curve is shown in Fig. 7.15. Over most of the frequency range up to 1.0 per-unit frequency the voltage is varied in direct proportion to the frequency. This type of characteristic provides a family of speed–torque curves with essentially constant breakdown torque as shown in Fig. 7.7.

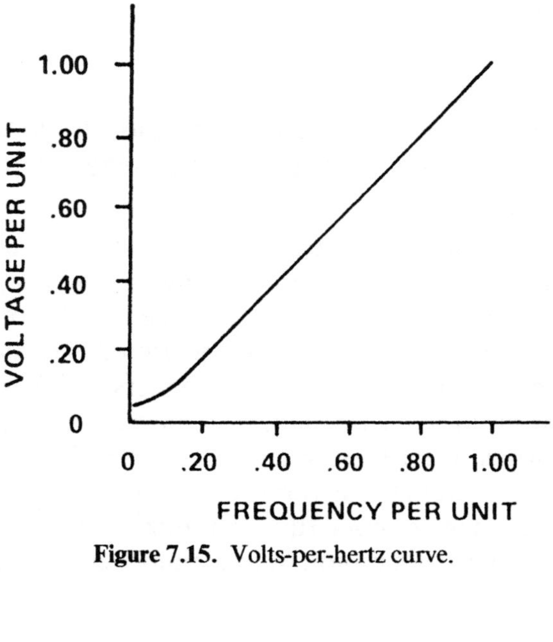


**Figure 7.15.** Volts-per-hertz curve.

At low frequencies, the primary resistance of the motor becomes a significant percentage of the motor's total impedance and the *IR* drop across the primary winding resistance results in a loss of torque unless some voltage boost is provided at the low end of the curve, hence the curved segment at the lower end of the volts-per-hertz curve in Fig. 7.15.

Starting must always be considered in the analysis and planning of any inverter application. This consideration is often neglected, much to the distress of the user when the installation is first tried and the inverter–motor combination is unable to start and accelerate the load. Motor users usually are not conditioned to be concerned about starting with fixed-frequency motors because most varieties of medium-size ac motors have starting torques ranging between 100% and 200% of full-load torque, which are adequate for most loads. It must be remembered that these starting torques are accompanied by inrush currents of approximately six times full-load current. With the relatively "infinite bus" supplying power to the induction motor in a typical fixed-frequency installation, the six times full-load inrush current creates no problem, and starting is simply not a matter of concern in the vast majority of cases.

However, with an inverter as the power supply, the situation is quite different. The rating of the thyristors or transistors in the inverter provides a limitation on the amount of current available for starting and accelerating the load. This limitation is usually between 100% and 150% of rated motor current if the motor and inverter are matched in size. This is far less than the 600% the user is accustomed to on fixed-frequency applications. Now, of course, with the inverter, the frequency is also reduced for starting and so the power factor under starting is improved which offsets, to some degree, the difference between the current values in the two situations. However, many loads do require considerably more than 100% of rated torque to break them away. If the load requires high breakaway torques for starting, then the inverter must be sized along with the motor to successfully accomplish this. Frequently, the determining factor is the starting rather than the full-load running requirement in specifying the size of both motor and inverter. Conveyors are one example of a type of load where this is often the case.

For fans and centrifugal pumps, starting is normally not a problem. Consequently, a linear volts-per-hertz relationship with little or no offset at the low-frequency end is usually acceptable and the motor–inverter combination does not have to be oversized for starting.

## 7.7. Harmonics and Their Effect on Motor Losses

Solid-state frequency converters accomplish their frequency synthesizing by switching one source of voltage on and off to produce a new frequency at the motor terminals. Inverters accomplish this by alternately switching the motor terminals between the positive and negative polarities of a dc bus. While this alternate switching between the positive and negative bus achieves the desired effect of controlling the frequency and voltage, and thus the motor's speed, it does so with the attendant disadvantage of increasing motor losses compared with a sine wave source of the same frequency and voltage.

The harmonic waveforms produced by an inverter can be explained by reference to a simplified schematic diagram of a variable-voltage input inverter shown in Fig. 7.16. The line-to-line voltage waveforms for this inverter are displayed in Fig. 7.17. To understand the switching required to achieve this waveform, consider an instant in time during which thyristors 1, 3, and 5 are conducting. This represents the region between 0 and 60° in Fig. 7.17. The motor phases A and C are connected to the positive dc bus and phase B is connected to the negative bus resulting in the line-to-line voltages shown. At the 60° transition point, thyristor 3 is commutated off and 6 goes into

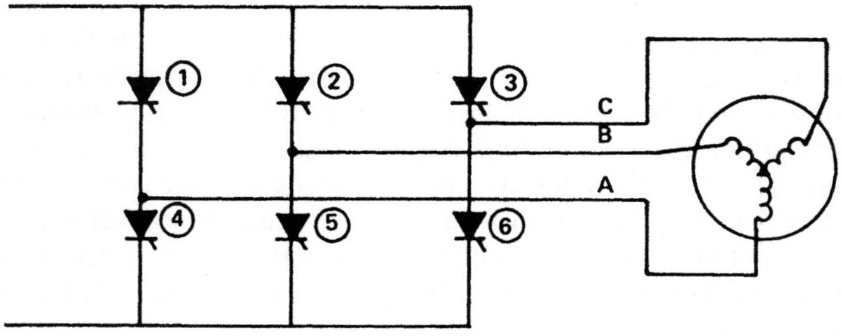


**Figure 7.16.** Simplified schematic diagram of variable-voltage input inverter and motor (feedback diodes and commutation circuits not shown).

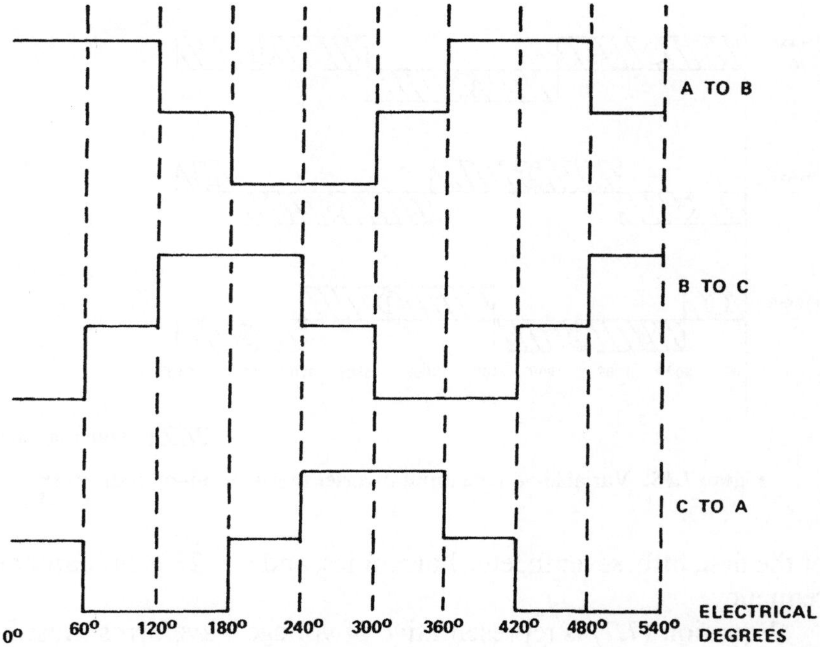


**Figure 7.17.** Line-to-line voltage waveforms of variable-voltage input inverter.

conduction producing the voltage pattern shown between 60 and 120°. Every 60 electrical degrees, one thyristor is commutated off and another one is turned on by a firing pulse to its gate. The conduction periods for each of the six thyristors in the inverter are shown in Fig. 7.18.

These square waves of voltage impressed on the motor terminals have a relatively high harmonic content and this affects motor losses. All of the line-to-line voltages are identical, only displaced from one another by 120°. These waveforms are periodic and have a finite number of discontinuities over a period which makes them amenable to analysis by Fourier series.[10] Using this technique, the voltage waveform can be described by

$$v(t) = \sqrt{2}[V_1 \sin\omega t + V_5 \sin 5\omega t + V_7 \sin 7\omega t + \cdots + V_k \sin k\omega t] \quad (7.7)$$

where $V_1$, $V_5$, $V_7$, . . . are the first, fifth, seventh, etc., rms voltages

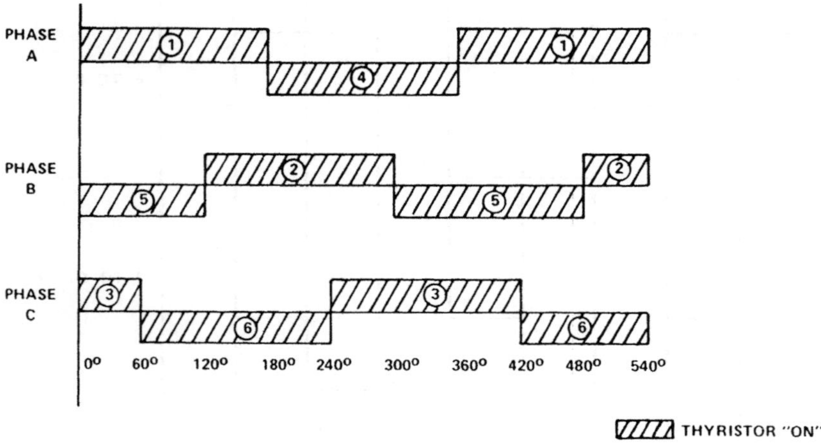


**Figure 7.18.** Variable-voltage input inverter thyristor on–off sequence.

for the first, fifth, seventh, etc., harmonics, and $\omega = 2\pi \times$ fundamental frequency.

Equation (7.7) is representative of voltage waveforms most frequently encountered with inverters applied to three-phase motors. It does not contain any even harmonics or harmonics which are divisible by 3.

The fifth, seventh, etc., harmonics are all time-varying harmonics which produce rotating magnetic fields in the air gap of the motor traveling at a speed of the fundamental field multiplied by the order of the harmonic. For example, in a four-pole, 60-Hz machine, the rotating magnetic field travels at 1800 rpm and the fifth harmonic field travels at

$$\text{Rotational speed of 5th time harmonic field} = 5 \times 1800$$
$$= 9000 \text{ rpm}$$

Another interesting feature of these rotating fields is that half of the harmonic fields rotate forward and half rotate backward as shown in Table 7.1.

An understanding of motor operation in the presence of harmonics can be obtained by considering the effects of each harmonic independently. Klingshirn and Jordan[11] have described this technique for the analysis of motors operating from inverters by considering the

**Table 7.1. Direction of Rotation of Time Harmonic Fields**

| Harmonic | Direction of rotation[a] |
|---|---|
| 1 | + |
| 5 | − |
| 7 | + |
| 11 | − |
| 13 | + |
| 17 | − |
| 19 | + |
| 23 | − |
| 25 | + |
| 29 | − |
| 31 | + |
| 35 | − |
| 37 | + |

[a] + indicates forward rotation; − indicates backward rotation.

motor as being excited from several independent generators, one representing each harmonic and all connected in series as shown in Fig. 7.19.

Since each harmonic current will be independent of all of the others, a series of independent equivalent circuits, one for each harmonic, can be used to calculate the complete steady-state performance of the motor when nonsinusoidal voltages are applied. This group of equivalent circuits is shown in Fig. 7.20. The results of the analysis carried out by this method describe the performance characteristics that can be expected of an inverter–motor combination. The details of such an analysis are given in Reference 11. A summary of

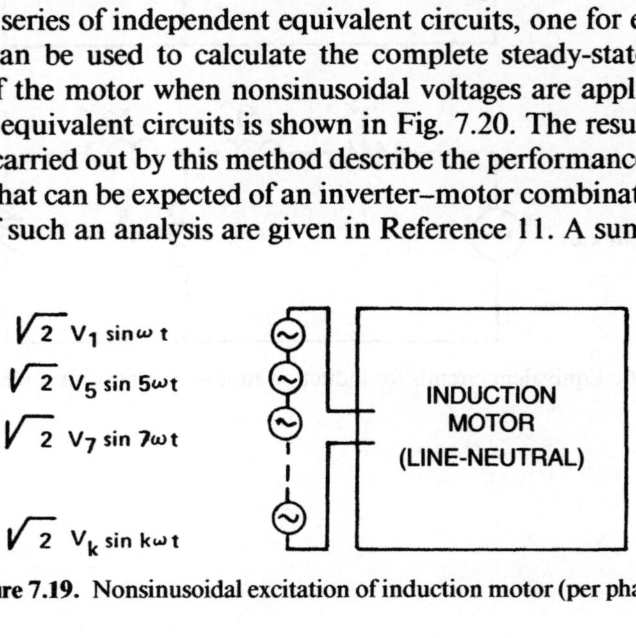


**Figure 7.19.** Nonsinusoidal excitation of induction motor (per phase).

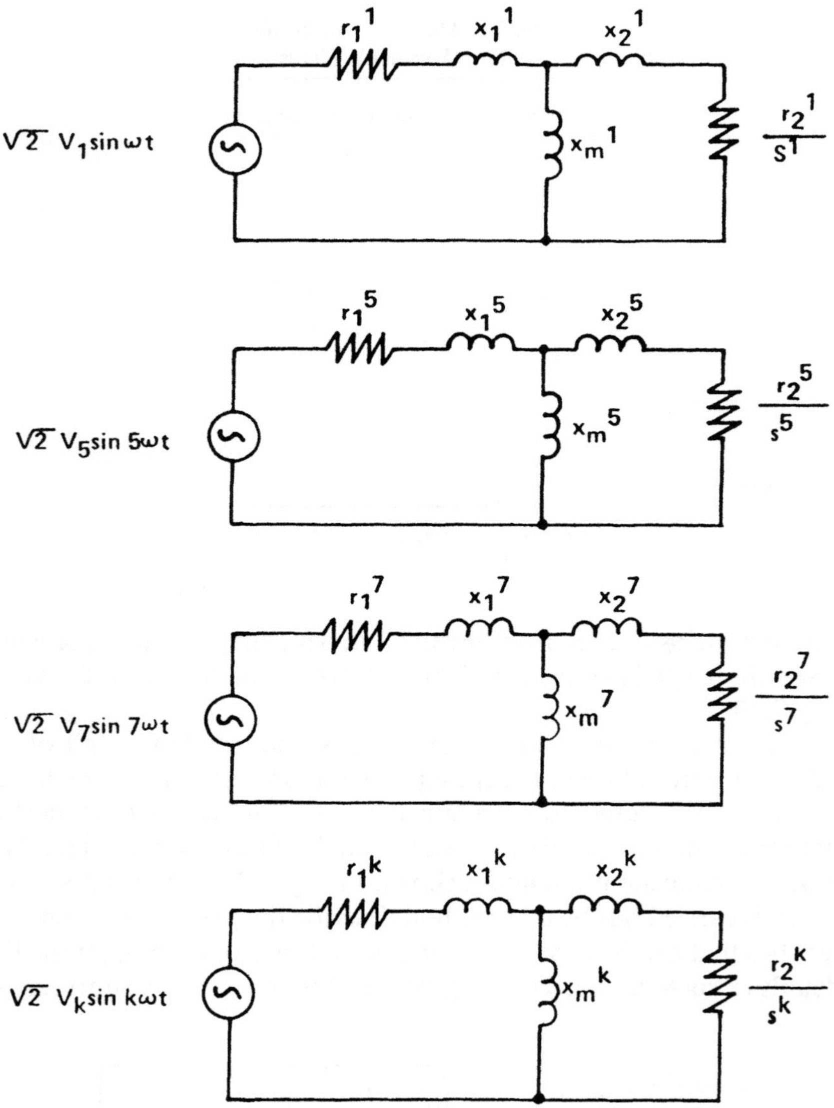


**Figure 7.20.** Equivalent circuits for induction motor with nonsinusoidal excitation.

the results is given here as it provides an understanding of the important factors to be considered in applying motors with inverters.

The harmonic content of a typical inverter voltage waveform is tabulated in Table 7.2. This "six-step" waveform is characteristic of variable-voltage input inverters. The percentage of each harmonic for the six-step waveform is given by

$$\text{Harmonic percentage} = \frac{100}{k}\,\% \tag{7.8}$$

where $k$ = order of the harmonic.

Many other waveforms are possible and are in use. Notable among these is the family of pulse-width-modulated inverter waveforms which are often tailored to suppress the 5th and 7th harmonics but, in exchange for this benefit, have large harmonic contents in the region of the carrier frequency. DeBuck[12] gives some typical pwm waveforms for 12 and 24 pulse-per-cycle modulation, which are tabulated in Table 7.3. The waveforms contain no 5th and 7th harmonics, but the presence of strong higher-order harmonics is evident.

Now each of these harmonics acts as a sinusoidal voltage impressed on its own equivalent circuit. The impedance of the circuits

**Table 7.2. Harmonic Content of Six-Step Waveform**

| Waveform harmonic | Voltage in percent of fundamental |
|---|---|
| 1 | 100.0 |
| 5 | 20.0 |
| 7 | 14.3 |
| 11 | 9.1 |
| 13 | 7.7 |
| 17 | 5.9 |
| 19 | 5.3 |
| 23 | 4.3 |
| 25 | 4.0 |
| 29 | 3.4 |
| 31 | 3.2 |
| 35 | 2.9 |
| 37 | 2.7 |
| 41 | 2.4 |
| 43 | 2.3 |
| 47 | 2.1 |
| 49 | 2.0 |

**Table 7.3. Harmonic Content of Pulse-Width-Modulated Waveforms[a]**

| | Voltage in percentage of fundamental | |
|---|---|---|
| Waveform harmonic | 12 pulses per cycle | 24 pulses per cycle |
| 1 | 100 | 100 |
| 5 | 0 | 0 |
| 7 | 0 | 0 |
| 11 | 40 | 0 |
| 13 | 40 | 0 |
| 17 | 4 | 0 |
| 19 | 11 | 0 |
| 23 | 13 | 40 |
| 25 | 13 | 40 |
| 29 | 18 | 0 |
| 31 | 9 | 0 |

[a] Data from DeBuck.[12]

appear to their respective voltage sources essentially as a motor at locked rotor. The reason for this is that the synchronous speeds of the harmonic fields are so much greater than the actual rotational speed of the motor that the values of $s^k$ are close to unity which approximates the locked-rotor condition. Therefore, the currents are limited mainly by the leakage reactances, $x_1^k$ and $x_2^k$. These currents contribute to additional $I^2r$ losses in the motor. The losses are increased by an additional amount related to the "skin" effect that exists in the rotors of most medium-size ac motors. Since the slips for the harmonics are large, approaching unity, the frequency of the rotor harmonic currents is also large and these currents crowd toward the top of the rotor cage thereby causing an increase in the effective rotor resistance and increased $I^2r$ losses in the rotor. These high-frequency currents and the associated high-frequency flux in the magnetic sections of the motor also cause substantial increase in iron losses. Taken in total, the increase in losses related to inverter harmonics is too much to be ignored. The added losses cause increased temperature rise in the motor and can materially affect motor life unless these losses are taken into account in selecting and applying the motor.

Usually, inverters are installed because operation below the motor's 60-Hz speed is desired. Unless the motor is provided with separate ventilation, the heat-dissipating ability of the machine decreases with speed because the motor's own fans deliver less cooling air. The combination of these two effects, increased harmonic losses and reduced heat-dissipating ability, must be considered in selecting the motor for adjustable-speed service.

These effects can be graphically displayed by a derating curve as shown in Fig. 7.21. The continuous torque carrying capability of the motor remains constant for some range of speeds below the rated speed and then drops off as the combined effect of harmonic losses and reduced ventilation begins to raise motor temperatures to unacceptable levels. Above rated speed, torque is inversely proportional to speed so that the motor delivers constant horsepower. In this region, the voltage output of the inverter remains constant as the frequency increases. Technically, the motor's torque capability decreases more rapidly than the curve would indicate since torque is proportional to $(\text{volts/hertz})^2$ but it is common practice to consider that, for a limited speed range above base speed, the motor can be operated at constant horsepower by utilizing some of the high breakdown torque capabilities that most induction motors have. Mechanical limitations may also limit operation above base speed. An investigation of the motor's mechanical limitations should be made to ensure that the machine

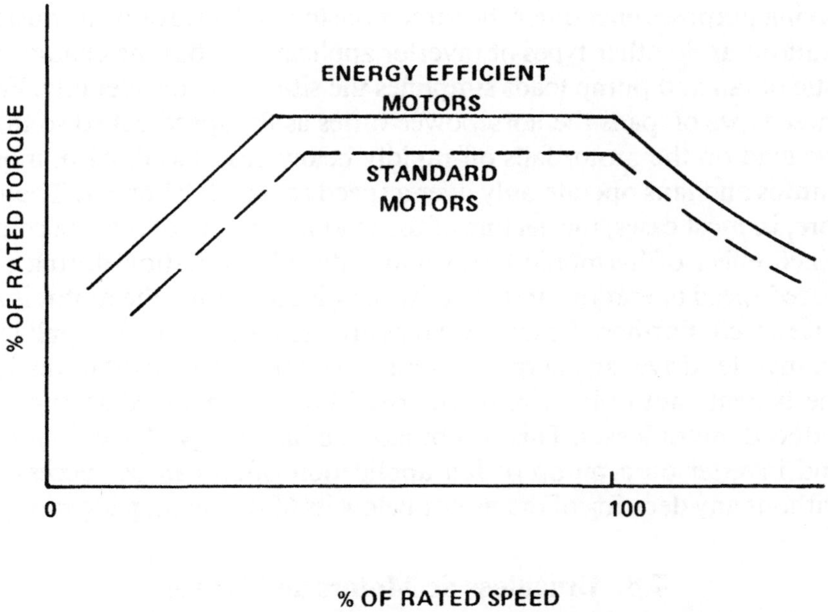


**Figure 7.21.** Derating curves for motors operating on inverter power supplies.

does not operate beyond its maximum safe operating speed or at a resonant speed condition.

Two curves are shown in Fig. 7.21, one for standard motors and one for energy-efficient motors. The derating penalty is less for an energy-efficient motor than for a standard motor. The energy-efficient motors contain more magnetic and conductor material than standard models so the harmonic currents and flux fields produce less additional losses. In addition, most energy-efficient motors operate on 60 Hz at less than their allowable temperature rise so there is more liberality to accommodate the additional heating resulting from inverter operation.

No simple, generalized rule can be stated for the derating of motors on general inverter applications. There are simply too many factors to consider, as is evident from the above discussion. Consequently, industry standards for motors operating from inverter power are not as comprehensive as the standards for fixed-frequency motors. The user must frequently seek from the manufacturer derating information which is specific to his motor, inverter, and application requirement.

Centrifugal pumps and fans which utilize inverters for energy-saving purposes encounter the same nonsinusoidal waveform considerations as do other types of inverter applications, but the characteristic of fan and pump loads simplifies the situation considerably. For these types of loads the horsepower varies as the speed cubed so that the load on the motor falls off rapidly below base speed. Also, most pumps and fans operate only over a speed range of 2:1 or 3:1. Therefore, in most cases, the decline of the load more than offsets the combined effect of harmonic losses and reduced ventilation during reduced speed operation, and the critical point for rating the motor is at base speed. Further, if energy savings are the motivation for applying an inverter drive, an energy-efficient motor should be used to realize the benefits not only of reduced speed but of the added savings of reduced motor losses. This combination of an energy-efficient motor and inverter on a pump or fan application often can be successful without any derating of the motor below its 60-Hz nameplate rating.

## 7.8. Brushless dc Motors and Drives

The term "brushless" applied to electric motors has been used to denote various motor architectures over the years. The "brushless dc motor" discussed here refers to an ac synchronous machine in which the stator, i.e., armature, windings are supplied through an electronic switching circuit. The rotor contains the motor field structure. The electronic switching circuit is an integral part of the "brushless dc" idea and serves the function of keeping the armature magnetic field locked in synchronism with the rotor field. The electronic switching is referred to as commutation because of the similarity in function to that provided by the mechanical commutator and brushes of a dc motor. The vast majority of brushless dc motors in service today have a three-phase ac stator winding and permanent magnets on the rotor. The discussion here will be limited to that configuration although other constructions are possible that still remain within the scope of the definition given here for "brushless dc."

Brushless dc motors derived their name because of their similarity to dc motors. The dc motor has long been the motor of choice in applications where speed, tension, position, etc., have to be controlled. The reason is the simplicity of control of a dc motor. The motor's torque is proportional to armature current, assuming a constant field and, therefore, the control of a single variable, armature current, accomplishes what is needed for a myriad of applications.

The brushless dc motor exhibits the same feature of producing torque which is proportional to current, at least to a good approximation. The brushless dc motor, however, is commutated electronically and does not have the disadvantage of requiring brushes.

The principle of operation of the brushless dc motor is that the rotor position is sensed and the magnetic field produced by the stator windings is electronically controlled to remain in synchronism with the magnetic field established by the rotor. The interaction of the control electronics and the motor will be explained with the aid of Fig. 7.22.

Figure 7.22 displays the three sinusoidal phase voltages $V_{an}$, $V_{bn}$, and $V_{cn}$ of a synchronous motor. The circuit diagram corresponding to this figure showing the power switching circuit for the brushless dc control and the synchronous motor is given in Fig. 7.23. The square waves shown below the phase voltages in Fig. 7.22 are the logic signals that turn the transistors $Q_1$ through $Q_6$ on and off. For example, $Q_1$ is turned on from 30° to 150° and again from 390° to 510°. From the logic switching sequence, it is clear that two phase windings are always conducting at any given instant in time. In particular, from 30° to 90° $V_{an}$ and $V_{bn}$ are connected to the positive and negative buses, respec-

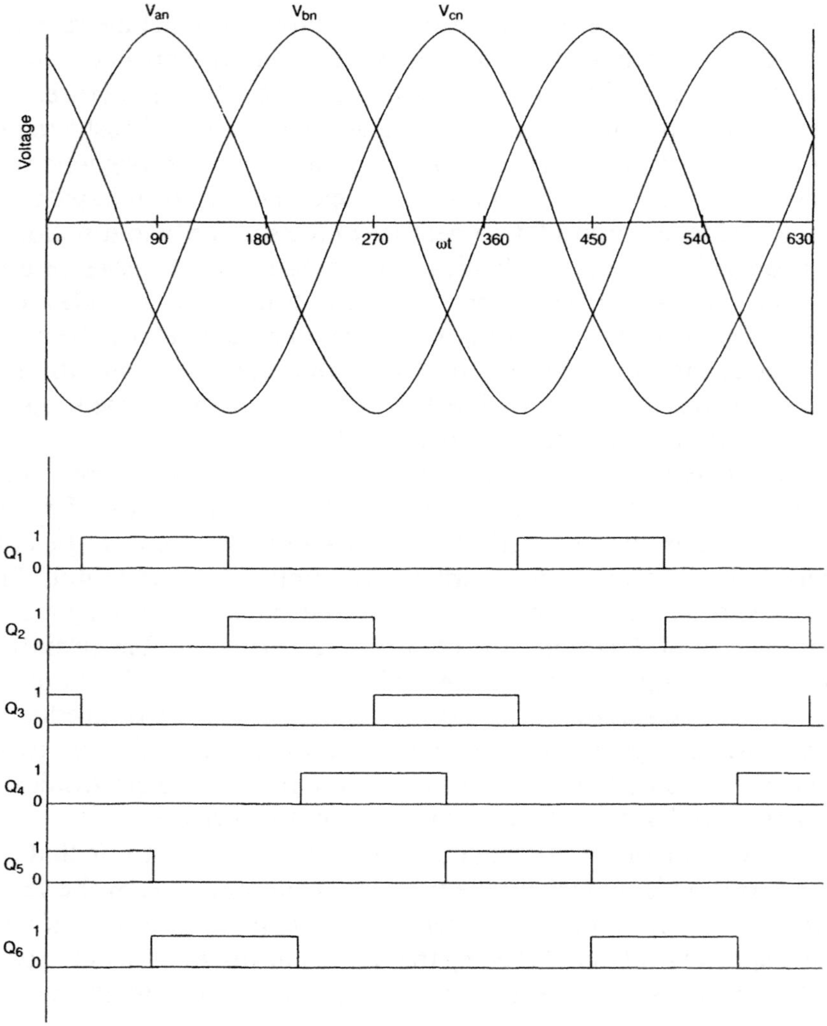


**Figure 7.22.** Brushless dc control switching diagram.

tively, and from 90° to 150°, $V_{an}$ and $V_{cn}$ are connected. This provides current into the motor during the peak part of the motor's counter emf cycle.

Of course, the current in the motor winding does not drop to zero instantaneously when a transistor switches off because of the circuit inductance. The diodes around each transistor provide a means for handling the inductive energy during the interval of time from transistor turn off to the time that the winding's current decays to zero.

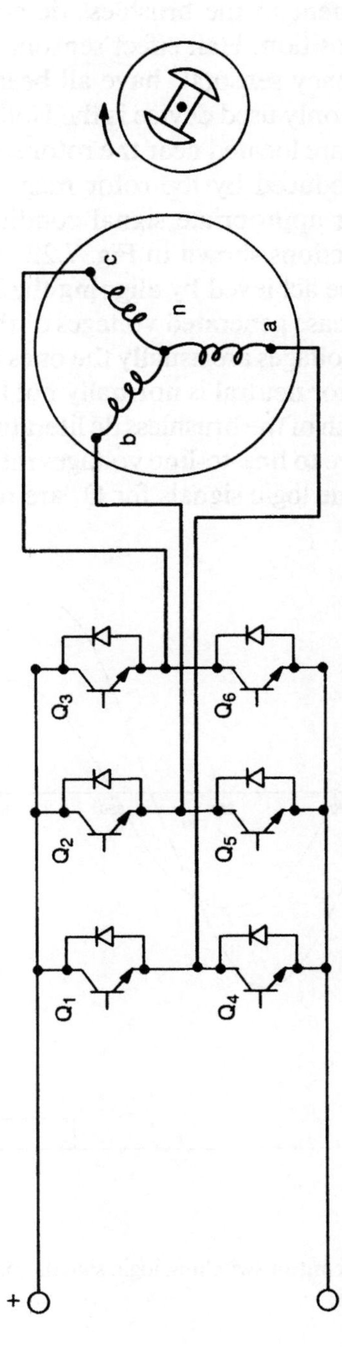


**Figure 7.23. Brushless dc motor and power electronics circuit.**

An essential element in the brushless dc control system is the sensing of the rotor position. Hall effect sensors, electro-optical sensors, and radio-frequency sensors[13] have all been used for this purpose. The most commonly used device is the Hall effect sensor.

The Hall devices are located near the rotor so that they can sense the magnetic field produced by the rotor magnets. The outputs of the Hall sensors, after appropriate signal conditioning, are used to generate the logic functions shown in Fig. 7.22. The proper Hall device positioning can be achieved by aligning the Hall device outputs relative to the three-phase generated voltages of the motor. However, the three line-to-line voltages are usually the ones available to be measured because the motor neutral is normally not brought out to a terminal. Therefore, much of the brushless dc literature displays the logic timing diagram relative to line-to-line voltages rather than phase voltages. Consequently, the logic signals for $Q_1$ are repeated in Fig. 7.24

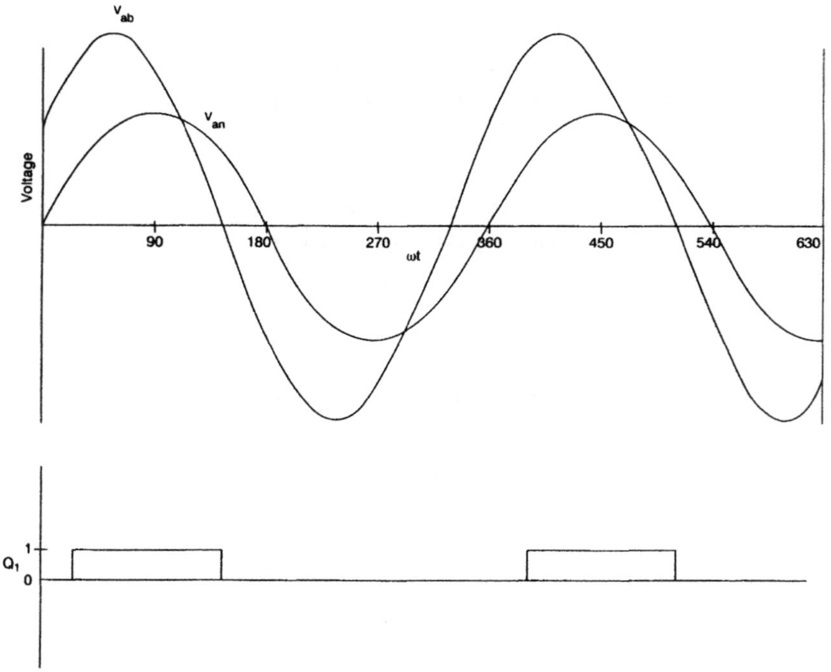


**Figure 7.24.** Brushless dc control switching logic signals compared with line-to-line and line-to-neutral voltages.

compared with the line voltage $V_{ab}$ and $V_{an}$ is included for reference. The 30° phase shift between $V_{ab}$ and $V_{an}$ is evident in the figure. The other five logic signals will have similar displacements relative to $V_{bc}$ and $V_{ca}$.

The preceding discussion has focused on one particular type of brushless dc drive. The motor voltage was assumed to have an ideal sinusoidal shape and the current was controlled to be rectangular pulses. The sinusoidal voltage/rectangular current drive requires some means of current control to regulate the motor torque. Recall that torque in this type of drive is proportional to motor current. A common technique is to pulse width modulate the active transistors.

Many brushless dc drives are constructed as described above but other configurations are certainly possible. Two additional architectures used in brushless dc drives are:

- Sinusoidal motor voltage and sinusoidal current
- Trapezoidal motor voltage and rectangular current pulse

Welch and Kaufman[14] compare the merits of various brushless dc configurations. The sinusoidal voltage/sinusoidal current system offers some performance advantages over the other types at the expense of some increased control complexity. However, the cost of control complexity becomes less important as integrated circuits become increasingly capable of embracing a large percentage of the control electronics.

Motors with trapezoidal voltage waveforms are more difficult to manufacture than the sinusoidal version so the sinusoidal voltage/rectangular current drive is the most common alternative to the sinusoidal voltage/sinusoidal current system.

Brushless dc drives have become so widely used, particularly in the fractional and lower horsepower sizes, that many integrated circuit manufacturers are providing a complete brushless control on a single chip. Silva[15] describes one such chip and there are many others.

The applications for brushless dc motors and drives range all the way from simple fan drives to high-response servo systems. Computer disk drives are one of the high-volume users of brushless dc. Brushless dc now occupies a major position in the arena of adjustable-speed drives.

## References

1. Schieman, R.G., Wilkes, E.A., and Jordan, H.E. Solid-state control of electric drives, *Proceedings IEEE,* Vol. 62, No. 12, December, 1974.
2. Phillips, K.P. Current source converter for Ac motor drives, *IEEE Transactions Industry Applications,* Vol. IA-8, No. 6, Nov./Dec. 1972.
3. Rosa, J. Utilization and rating of machine commutated inverter-synchronous machine drives, *IEEE Industry Application Society Meeting,* New York, 1978.
4. Bose, B.K. *Power Electronics and Ac Drives.* Englewood Cliffs, N.J.: Prentice–Hall, 1986.
5. Mokrytzki, B. Pulse modulated inverters for Ac motor drives, *IEEE Transactions on Industry and General Applications,* Vol. IGA-3, Nov./Dec., 1967.
6. Boldea, I., and Nasar, S.A. *Vector Control of Ac Drives.* Boca Raton, Fla.: CRC Press, 1992.
7. Vas, P. *Vector Control of Ac Machines.* London: Oxford University Press (Clarendon), 1990.
8. NEMA Standards Publication No. MG1, "Motors and Generators." Washington, D.C.: National Electrical Manufacturers Association, 1993.
9. Tsivitse, P.J., and Klingshirn, E.A. Optimum voltage and frequency for polyphase induction motors operating with variable frequency power supplies, *IEEE Industry and General Applications Group Conference Record,* 1970.
10. Wylie, C.R., Jr. *Advanced Engineering Mathematics.* New York: McGraw–Hill Book Co., Inc., 1960.
11. Klingshirn, E.A., and Jordan, H.E. Polyphase induction motor performance and losses on non-sinusoidal voltage sources," *IEEE Power Apparatus and Systems Transactions,* Vol. PAS87, 1968.
12. DeBuck, F.G.G. Losses and parasitic torques in electric motors subjected to PWM waveforms, *IEEE Industry Applications Transactions,* Vol. IA-15, Jan./Feb., 1979.
13. "DC Motors, Speed Controls, Servo Systems," an engineering handbook by Electro-Craft Corp., 1980.
14. Welch, R., and Kaufman, G. Performance of a three phase sinusoidal Ke motor using a sinusoidal amplifier and a trapezoidal amplifier, *Third Annual Motion Control Technology Conference,* Boston, March, 1991.
15. Silva, C. Brushless DC motors get a controller IC that replaces complex circuits, *Electronic Design,* September 19, 1985.

# 8

# ac Motor Control and Protection: Starting, Running Protection, and Surge Protection

## 8.1. Overview of Starting Requirements and Controls

Starting methods for ac motors range across a broad spectrum from across-the-line starting to various types of reduced voltage starters; more recently, electronically controlled starting is being used. When rated voltage is applied to a motor at standstill, the inrush current is several times the rated load current. For motors in the medium ac size, the starting torque accompanying this high inrush current generally exceeds 100% of full-load torque, and for smaller horsepower sizes, starting torques of 200% or more are common. For most applications, these starting torques are adequate to start the load and, if the power system supplying the motor is not adversely affected by the inrush current, across-the-line starting can be used. However, if the power supply suffers a severe voltage dip when required to supply the motor's inrush current, then an alternate type of motor starting control may be needed. Light flicker is a common symptom of voltage reductions that occur during motor starting. This is often tolerable, but if the voltage reduces to such a low level that the motor will not start the load, then some corrective action needs to be taken. Computers and production equipment may also be susceptible to voltage dips and may malfunction. If any of these problems create unacceptable conditions, then reduced voltage starting is indicated.

The starting phenomenon takes place in two phases, a transient and a steady-state phase. An inrush current waveform is displayed in Fig. 8.1. At the instant the circuit is energized, the current is composed of two components, a transient and a steady-state component. The net current is the sum of these two components.

A transient component occurs in any inductive circuit which is suddenly connected to a sinusoidal voltage source. It is the component necessary to satisfy the difference between the steady-state component and the total current required by the circuit conditions at the instant of switch closure. The transient component dies away exponentially and only the steady-state component remains. The first peak value of current may be approximately 2.8 times the rms value of the motor's locked-rotor current. Whether or not the transient peak reaches this value depends on the instant of closure and, in general, the transient peaks will be different in each of the three phases.

Usually, all references to inrush currents are based on the steady-state component. This is the value quoted by motor manufacturers as "locked rotor" or "inrush" current. Occasionally, the first current peak, $I_T$ in Fig. 8.1, will cause circuit breakers to trip even though the motor's locked-rotor current is within the value that the circuit

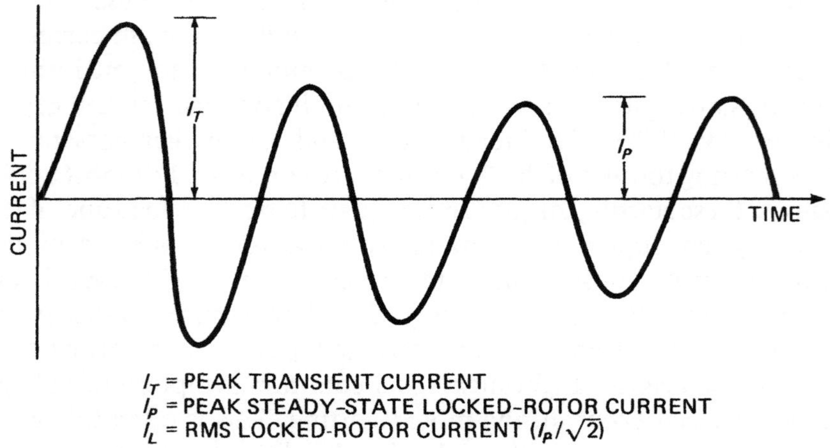


**Figure 8.1.** Current versus time for across-the-line starting of an induction motor from a constant voltage source.

breaker is rated to handle. The instantaneous trip setting can be adjusted upward to avoid this problem and this is permitted by the National Electrical Code[5] up to some maximum value which is a multiple of the motor's rated current.

The starting performance and the application of motor controls are generally the same for both energy-efficient motors and standard motors so long as both meet NEMA Design B characteristics.

## 8.2. Across-the-Line Starting

The term "across-the-line" starting refers to the process of connecting the motor directly to the power line, and this is usually done by closing a magnetically operated contactor. Across-the-line starting is in contrast to reduced voltage starting, discussed in later sections,

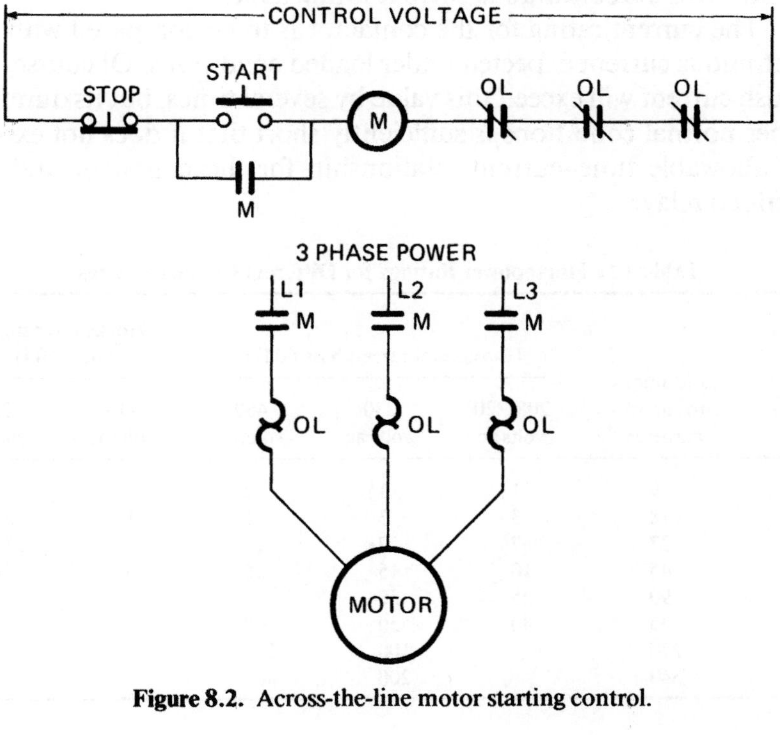


**Figure 8.2.** Across-the-line motor starting control.

where there are one or more intervening steps of reduced voltage application to the motor intended to reduce the inrush current.

Across-the-line starters consist of a contactor, the magnetic control for the contactor, and overload relays. The contactor is closed and maintained in a closed position by a holding coil, which is excited from a control voltage source, usually 115 V, 60 Hz. Wired in series with the holding coil are the start and stop push buttons and the contacts of three overload relays. The current sensing elements of these overload relays are in series with each of the three power leads to the motor. If the current of the motor exceeds a predetermined rms value and time, the overload relay opens, disconnecting the motor from the line.

A wiring diagram for an across-the-line starter is shown in Fig. 8.2. Standard sizes and continuous-current carrying capacities have been established for general-purpose contactors. Table 8.1 is a tabulation of contactor sizes 00 through 6 together with the continuous-current rating of each size, and the horsepower rating which corresponds with the continuous current rating table.

The current rating for the contactor is to be compared with the continuous current expected under loaded conditions. Of course, the inrush current will exceed this value by several times, but its duration under normal conditions is sufficiently short that it does not exceed the allowable time–current relationship for the contactor and the overload relay.

**Table 8.1. Horsepower Ratings for Different Contactor Sizes**

| | | Horsepower three-phase 60 Hz | | | Horsepower single-phase 60 Hz | |
|---|---|---|---|---|---|---|
| Size | Continuous current, rms amperes | 208/220 volts ac | 230 volts ac | 460 volts ac | 115 volts ac | 230 volts ac |
| 00 | 9 | $1\frac{1}{2}$ | $1\frac{1}{2}$ | 2 | $\frac{1}{3}$ | 1 |
| 0 | 18 | 3 | 3 | 5 | 1 | 2 |
| 1 | 27 | $7\frac{1}{2}$ | $7\frac{1}{2}$ | 10 | 2 | 3 |
| 2 | 45 | 10 | 15 | 25 | 3 | $7\frac{1}{2}$ |
| 3 | 90 | 25 | 30 | 50 | | |
| 4 | 135 | 40 | 50 | 100 | | |
| 5 | 270 | | 100 | 200 | | |
| 6 | 540 | | 200 | 400 | | |

## 8.3. Primary-Impedance Starting

Primary-impedance starting achieves inrush current reduction by inserting three impedances, one in each power line, in series with the motor. A circuit diagram for a resistance starter is shown in Fig. 8.3. Resistance starters are available through most of the medium horsepower range, up to approximately 200 hp, and reactors are used for impedance starting of large ac motors. The "Start" contacts, "S" in Fig. 8.3, close first and at some later time, the "Run" contacts close, bypassing the series resistors. After the "Run" contacts are closed, the "S" contacts are open. A timer is used to control the point at which the transition between Start and Run takes place.

This type of transition between Start and Run is referred to as a closed transition because the motor remains energized continuously from the time the start is initiated. Other types of reduced voltage starters remove the excitation from the motor for a brief period of time during the transition from one voltage step to the next. Closed transitions minimize the transient inrush and transient torque peaks and thereby reduce shock loads on the power supply system and stress on the motor winding.

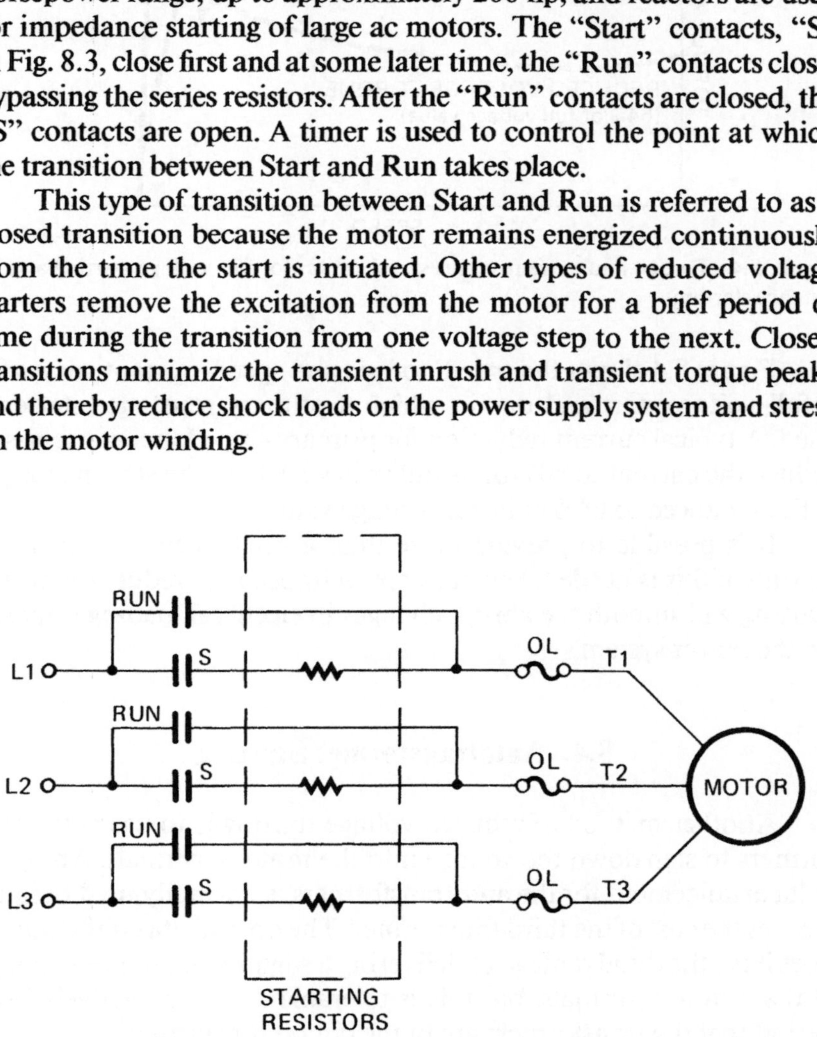


**Figure 8.3.** Primary resistance starting of induction motor.

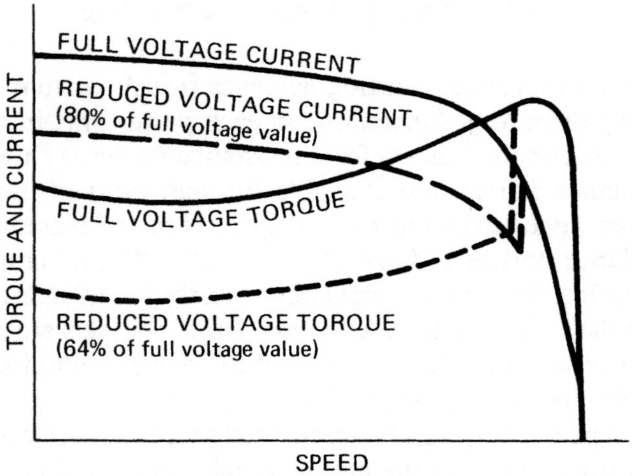


**Figure 8.4.** Torque and current versus speed curves for full voltage starting and resistance starting.

Figure 8.4 shows the speed–torque and speed–current curves for a full voltage start and one in which a primary-resistance starter is used. A typical current reduction for primary-impedance starters is to reduce the current to 80% of its full voltage value. The starting torque is then reduced to 64% of its full voltage value.

It is possible to provide more than a single step of impedance starting if this is needed. Several steps of impedance reduction during starting will smooth the abrupt changes in electrical loading imposed on the power system.

## 8.4. Autotransformer Starting

Another method of reduced voltage starting is to use autotransformers to step down the voltage into the motor terminals. An open-delta arrangement for the autotransformers is normally used because it saves the cost of the third transformer. The open-delta configuration does have the disadvantage of delivering a slightly unbalanced voltage at the output terminals, but this is tolerable over the relatively brief period that the transformers are in the circuit for starting.

Two types of autotransformer starters are available, open and closed transition. These are shown in Figs. 8.5 and 8.6, respectively. The closed transition system reduces the step voltage changes applied to the motor considerably over the step changes with an open transition system. In the closed transition arrangement, there are actually two steps of voltage during the accelerating period. This is shown in Fig. 8.7 which displays curves of current and torque versus speed for a closed transition, autotransformer starter. Two or three selections of voltage reduction are usually provided with autotransformer starters. The selection is made by connecting to the proper output terminals of the transformer. Table 8.2 tabulates three of the commonly used voltage output percentages and the inrush currents and torques expressed as a percentage of the value which would be realized if full voltage was applied. Note that the current taken from the power supply system for an autotransformer starter is reduced by the square of the voltage reduction tap rather than the first power. The current into the motor is, of course, reduced by the same percentage as the voltage tap but, because of the autotransformer effect, the current taken from the power line is reduced by the square of the voltage reduction per-

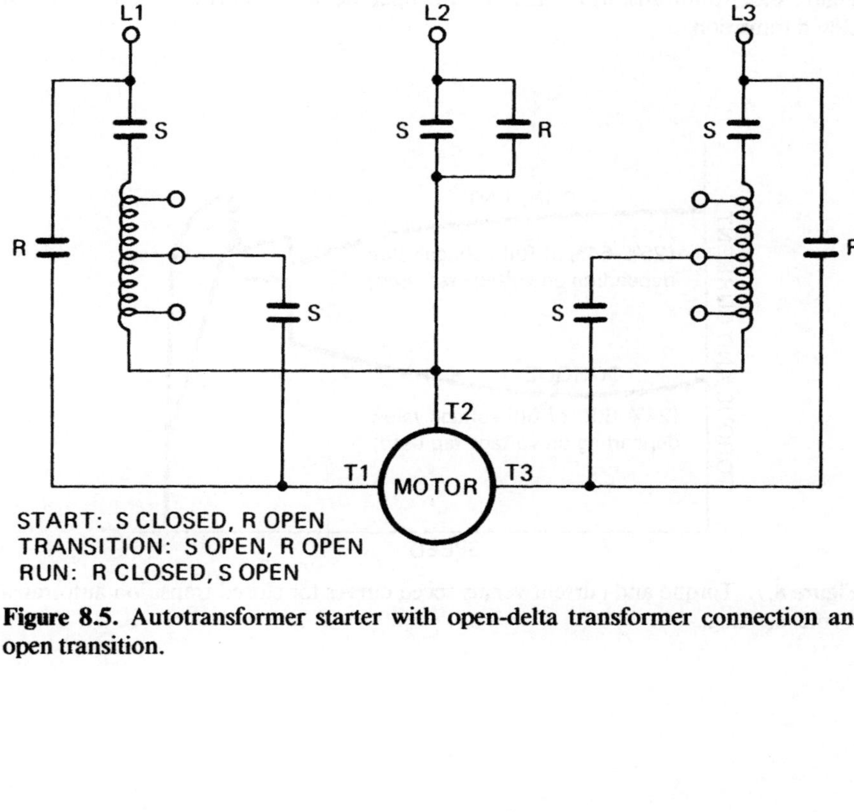


**Figure 8.5.** Autotransformer starter with open-delta transformer connection and open transition.

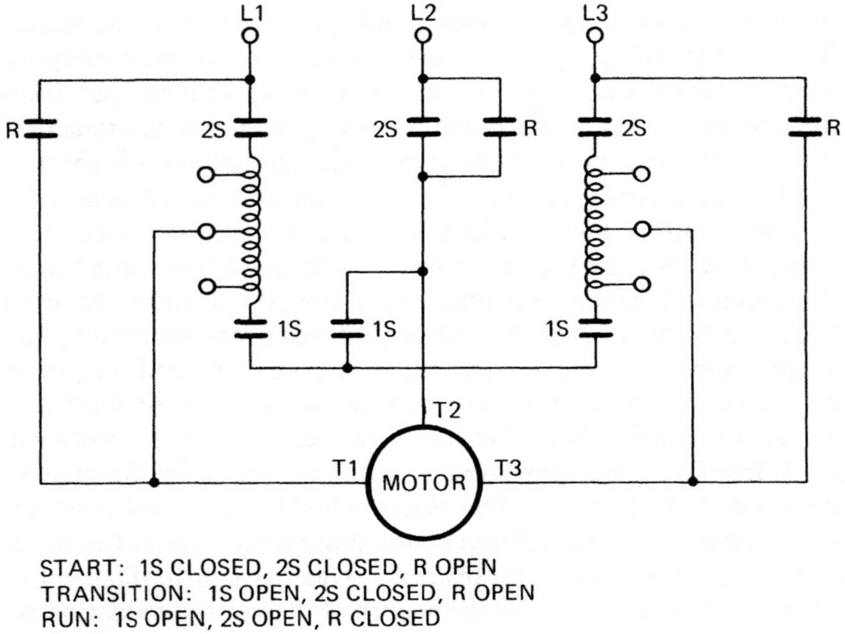


**Figure 8.6.** Autotransformer starter with open-delta transformer connection and closed transition.

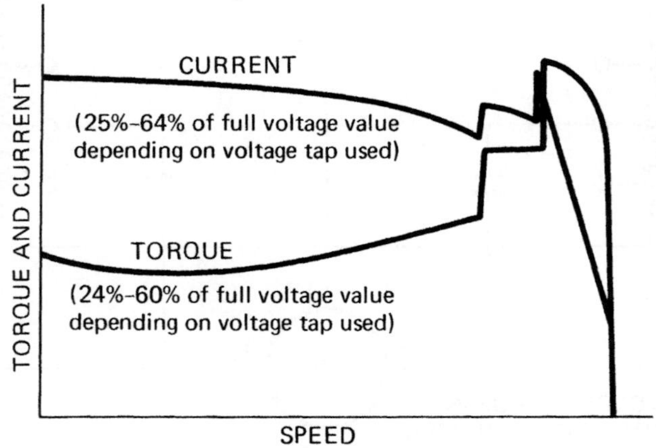


**Figure 8.7.** Torque and current versus speed curves for closed transition autotransformer starting.

**Table 8.2. Locked-Rotor Currents and Torques for Autotransformer Starting**

| Autotransformer tap | % of locked-rotor current | % of locked-rotor torque |
|---|---|---|
| 50% | 25% | 24% |
| 65% | 42% | 40% |
| 80% | 64% | 60% |

centage. Torque is also reduced by approximately the square of the voltage reduction. Actually, the torque reduction is somewhat greater than the second power of voltage because of the impedance drop in the transformer and the loss of torque resulting from unbalanced voltage. These relationships are evident in Table 8.2.

Autotransformer starters are quite effective in reducing current taken from the power line. However, the transformers increase the cost of the starter and this type of reduced voltage starter is more expensive than some of the other types discussed in this chapter.

## 8.5. Wye–Delta Starting

Wye–delta starting is a technique which requires no transformers, resistors, or reactors to be inserted in series with the motor. However, this type of starter requires a six-lead motor so that the winding can be switched from wye to delta during the starting interval. The motor is started with the windings connected in wye, and then switched to a delta connection for running. This transition is usually controlled by a timer.

When the motor is connected in wye, the voltage across each motor phase winding is 58% of the value that it assumes after the winding has been switched to delta. The line current during the wye start is approximately 33% of the value it would be if the motor were started on the delta connection and the torque with the wye connection is 33% of the delta starting torque. The starting and accelerating torques with this type of starter are relatively low and therefore wye–delta starting cannot be applied to loads requiring high breakaway torques. A common application for the wye–delta starter is in the starting and accelerating of compressor-type loads since, for these types of loads, the breakaway torque is usually small and the load torque follows a squared curve as the speed increases. If the compressor is started unloaded, the motor may be capable of accelerating it all the way to full speed on the wye connection. However, if the load torque and motor torque curves intersect at any point below operating speed, then the timer must be set so that the transition to the delta connection takes place near the time when this intersection occurs. If not, the motor will operate at the point of intersection of the two curves and will overheat.

Wye–delta starters are available in both open and closed transition varieties, and the diagrams for these types of starters are displayed in Figs. 8.8 and 8.9. Speed, current, and torque curves are shown in Fig. 8.10. In this figure the transition between windings occurs in the region just below breakdown torque. This is normally the critical re-

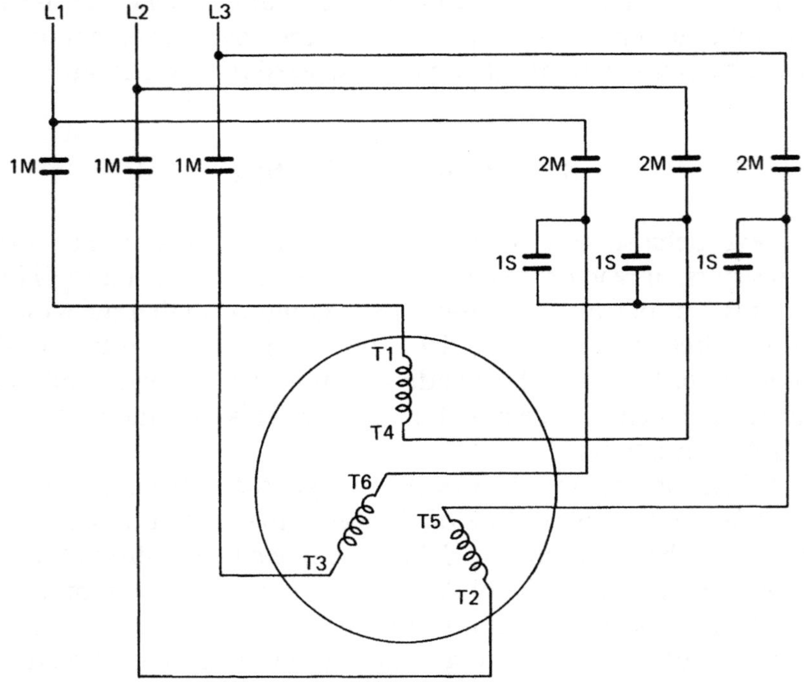


| FUNCTION | CONTACTORS 1M | 2M | 1S |
|---|---|---|---|
| Start | CLOSED | OPEN | CLOSED |
| Transition | CLOSED | OPEN | OPEN |
| Run | CLOSED | CLOSED | OPEN |

**Figure 8.8.** Wye–delta starter with open transition.

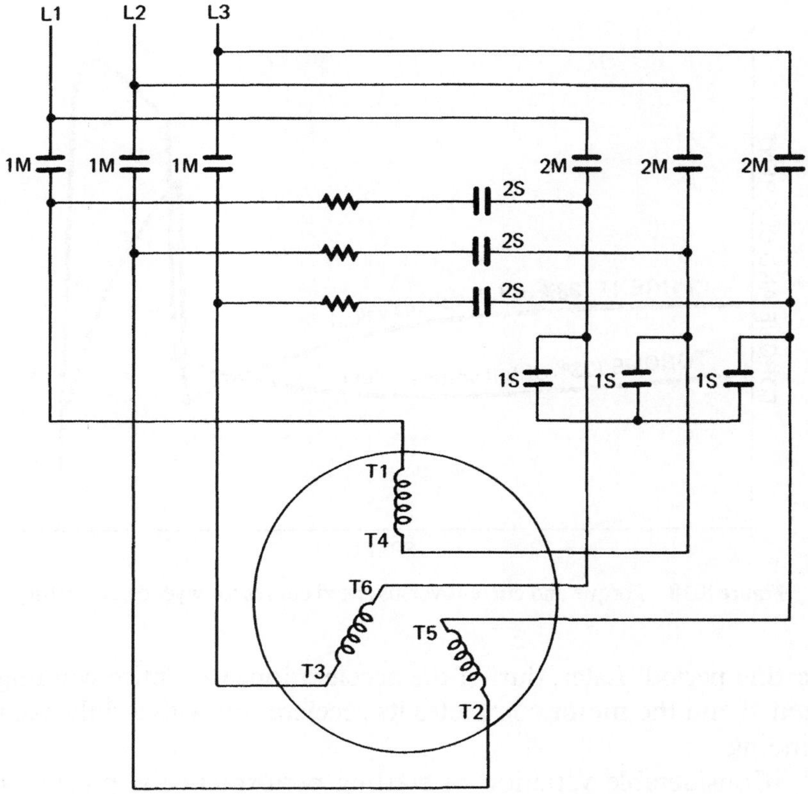


| FUNCTION | CONTACTORS 1M | 2M | 1S | 2S |
|---|---|---|---|---|
| Start | CLOSED | OPEN | CLOSED | OPEN |
| Transition 1 | CLOSED | OPEN | CLOSED | CLOSED |
| Transition 2 | CLOSED | OPEN | OPEN | CLOSED |
| Transition 3 | CLOSED | CLOSED | OPEN | CLOSED |
| Run | CLOSED | CLOSED | OPEN | OPEN |

**Figure 8.9.** Wye–delta starter with closed transition.

gion for accelerating fan types of loads and, therefore, the point at which switching to the delta should take place.

## 8.6. Part-Winding Starting

Part-winding starting achieves its inrush current reduction by exciting only part of the motor winding during the first phase of the

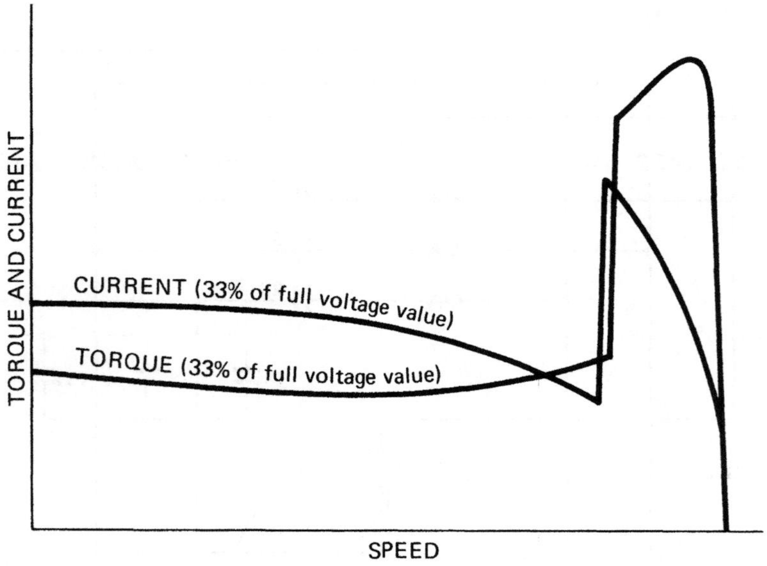


**Figure 8.10.** Torque and current versus speed curves for wye–delta starting.

starting period. Later, during the acceleration, the entire winding is excited and the motor completes its acceleration with a fully excited winding.

Considerable variation of starting performance is possible depending on the motor winding configuration used during the part winding operation. Courtin[2] reports test results on ten different configurations which are possible for one four-pole design. Each of these produces a different acceleration characteristic.

The excitation of only half of the winding results in a nonsinusoidal magnetic field in the motor. In fact, the air-gap flux of the motor consists of several different fields rotating at different speeds. A winding connection commonly used for part-winding start motors results in a strong, forward rotating field with two poles more than the fundamental field in addition to the fundamental field itself. For example, a four-pole motor operating on the part-winding connection would have both four- and six-pole fields present simultaneously. As the motor accelerates through the synchronous speed of the six-pole field, 1200 rpm with 60-Hz excitation, a strong breaking torque is produced causing a cusp in the motor's speed–torque curve. This cusp may be so severe that the motor's net torque is negative through this segment of the speed range. The acceleration will, of course, be stopped if the net motor torque goes negative. Even if the net motor torque remains positive, the motor may not be able to accelerate through this region if there is significant retarding torque imposed by the load. It is important, therefore, that the starter's timer be set to excite the full winding at approximately the time that the motor reaches a speed region where reduced torque is present. If this is done, prolonged operation at a single speed during the accelerating period will be avoided. If the timer is not properly set, the motor may overheat.

A common application for part-winding starters is to use them with a nine-lead, dual-voltage motor. This must be done with the motor connected to operate on the lower of the two rated voltages. It is advisable if one intends to use this method of starting, that verification of the motor's suitability for part-winding start operation be obtained, as not all dual-voltage motors will perform satisfactorily.

Figure 8.11 displays the connection diagram and Fig. 8.12 shows the speed–torque and speed–current curves for part-winding starting. The starting current is reduced to approximately 67% of its rated voltage value and the starting torque is 45% of the motor's full-voltage

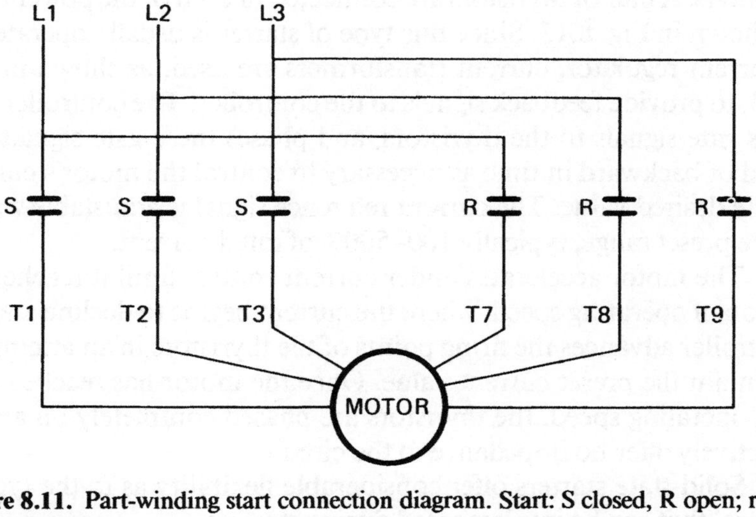


**Figure 8.11.** Part-winding start connection diagram. Start: S closed, R open; run: S closed, R closed.

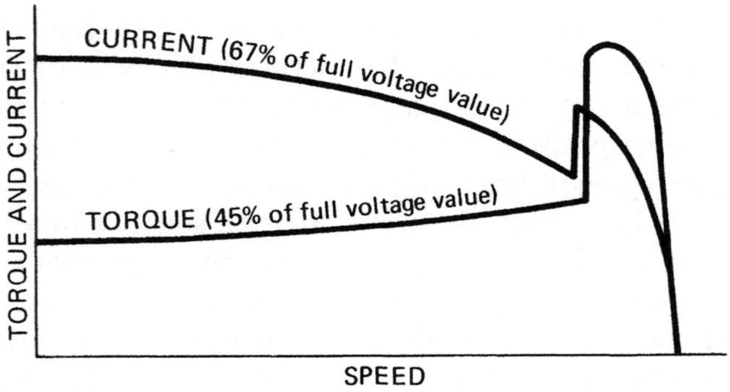


**Figure 8.12.** Torque and current versus speed curves for part-winding starting.

torque during the interval in which the motor is excited on part winding.

## 8.7. Solid-State, Electronic Starting

From the standpoint of minimizing stresses on the electrical system caused by inrush current, solid-state starting is one of the best methods. A pair of thyristors are connected in each of the power lines, as shown in Fig. 8.13. Since this type of starter is usually operated as a current regulator, current transformers are used, as shown in Fig. 8.13, to provide feedback signals to the controller. The controller supplies gate signals to the thyristors, and phases these gate signals forward or backward in time as necessary to control the motor's current to the desired value. The current reference signal is adjustable within some preset range, typically 100–500% of rated current.

The motor accelerates under current control until it reaches the region of operating speed, where the current begins to decline and the controller advances the firing points of the thyristors in an attempt to maintain the preset current value. Once the motor has reached normal operating speed, the thyristors are phased completely on and so effectively offer no impedance to the circuit.

Solid-state starters offer considerable flexibility as to the type of control that can be implemented during starting.

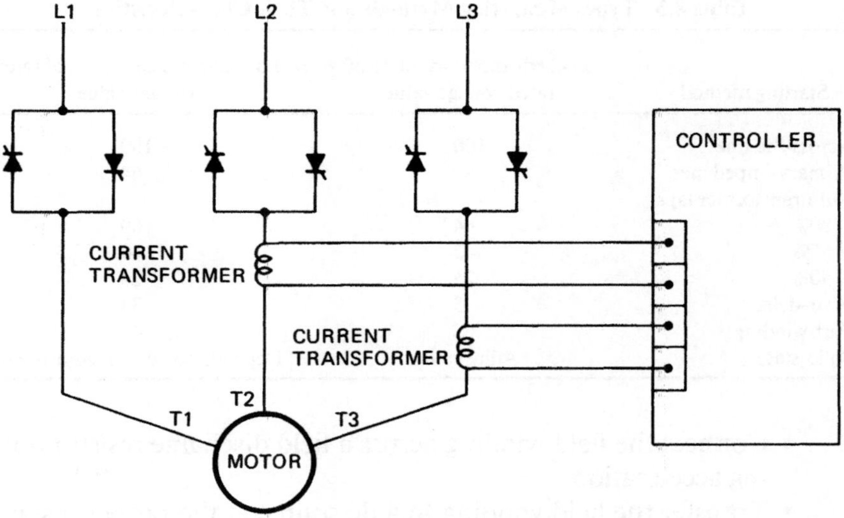


**Figure 8.13.** Solid-state electronic starter.

- Current ramp—The current is ramped up to its preset value and maintained constant throughout the accelerating period.
- Speed ramp—Current is controlled to provide a speed versus time profile. A speed feedback signal from a tachometer is required for this type of control.

Other types of controls are, of course, also possible because of the inherent flexibility using an electronic controller.

Table 8.3 presents a summary of the various starting methods and a comparison of their performance characteristics.

## 8.8. Synchronous Motor Starting

A synchronous motor starter must combine the functions of connecting the motor's armature across the line with the function of applying the field excitation. Synchronous motors start and accelerate as induction motors using a cage winding on the rotor as the short-circuited secondary. As the motor nears synchronous speed, the field excitation is applied and the motor pulls into synchronism. The synchronizing control must perform two functions:

**Table 8.3. Types of Starting Methods and Their Characteristics**

| Starting method | Locked-rotor current, % of rated voltage value | Locked-rotor torque, % of rated voltage value |
|---|---|---|
| Across-the-line | 100 | 100 |
| Primary-impedance | 80 | 64 |
| Autotransformer taps | | |
| 80% | 64 | 60 |
| 65% | 42 | 40 |
| 50% | 25 | 24 |
| Wye–delta | 33 | 33 |
| Part-winding | 67 | 45 |
| Solid-state | Adjustable | Depends on current adjustment |

- Connect the field winding across a field discharge resistor during acceleration
- Transfer the field winding to a dc source at the proper instant in time

A field discharge resistor is necessary during the accelerating period to avoid high voltages from appearing across the motor field. The transfer of the field from the discharge resistor to the dc supply ideally should take place at an instant when the rotating field pole's polarity is aligned with the armature magnetic field, so that the rotor has to accelerate through only a small angle to achieve synchronism. If the alignment is wrong, then the rotor will slip one or more poles in its attempt to synchronize. This can create large transient torques at the motor shaft and large pulsating currents in the electrical system.

Varying degrees of sophistication are incorporated into synchronizing control circuits. One method for sensing the correct instant to disconnect the discharge resistor and apply the dc field current is by using a polarized field relay. The operation of this type of relay is shown in Fig. 8.14, and a control circuit for starting a synchronous motor using the polarized field relay is shown in Fig. 8.15.

Referring to Fig. 8.14, the ac voltage induced in the motor field during starting causes current to pass through the ac coil of the polarized field relay. The current produces sufficient flux to pull the contact PFR open and hold it open during the accelerating period. As the motor nears synchronous speed, both the frequency and voltage induced in the field winding decline. Most of the ac current in the motor

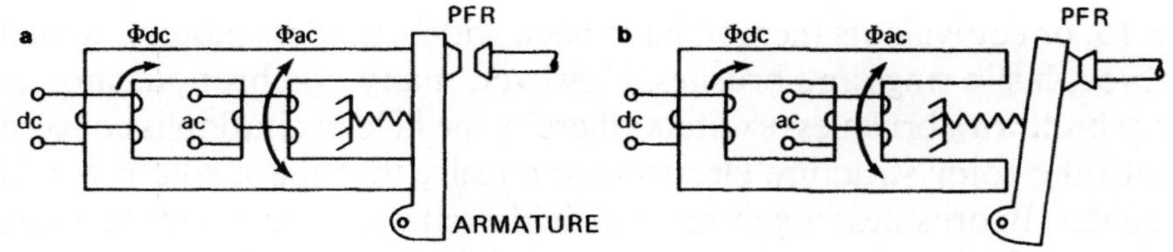


**Figure 8.14.** Polarized field relay. (a) Starting–ac flux (Φac) holds contact PFR open. (b) Synchronizing–net flux (Φac–Φdc) drops to a low value and PFR closes. Polarizing flux Φdc causes PFR to close at correct polarity.

field circuit passes through the inductance $X_l$, which shunts the PFR relay as shown in Fig. 8.15, and the ac flux in the PFR relay drops to a low value. This causes the contact PFR to return to a closed condition. The closure occurs on a half cycle when the ac flux and the polarized dc flux are in phase opposition, thus establishing the correct polarity relationship for the field excitation to be applied.

The electromechanical relay system shown in Figs. 8.14 and

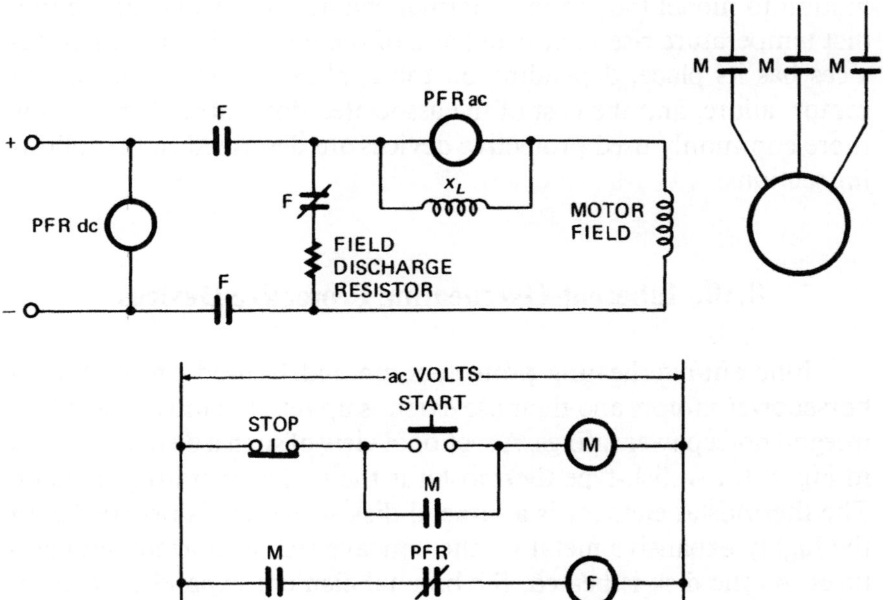


**Figure 8.15.** Simplified control circuit for synchronous motor starting using a polarized field relay.

8.15, or equivalents thereof, have been widely used in motors excited through slip rings and brushes. However, many synchronous motors are built with brushless exciters whereby the field excitation is coupled into the rotor structure electromagnetically through a rotating field exciter. In brushless machines, the field excitation control must rotate as part of the rotor structure, and electromechanical relays are usually not used. The function of transferring the field circuit from a discharge resistor to a dc source is implemented with solid-state electronics for brushless synchronous motors.

## 8.9. Motor Overtemperature Protection Methods

Motor overtemperature protection embraces a wide variety of devices and techniques. These range all the way from relatively simple current relays which sense motor current and trip on overcurrent, to solid-state devices which receive multiple inputs and use this information to model the motor's thermal characteristics in order to predict temperature rise in critical parts of the motor. Each of these devices has its place, depending on the application need, the cost of motor failure, and the cost of the associated downtime. Some of the more commonly used protective devices are described in the following sections.

## 8.10. Inherent-Overheating Protective Devices

Inherent-overheating protectors are widely used on fractional-horsepower motors and their use extends up into the lower end of the integral horsepower ratings. A sectional view of such a device is shown in Fig. 8.16. A disk-type thermostat is the major operating element. The thermostat element is a bimetal disk which is dish-shaped with the highly expansive metal on the concave side at normal temperatures. As the disk is heated, the bimetal element expands, but is restrained by the less expansive metal on the outside. At some critical value, the disk snaps into a convex shape, opening the electrical contacts which interrupt current to the motor. The motor current also

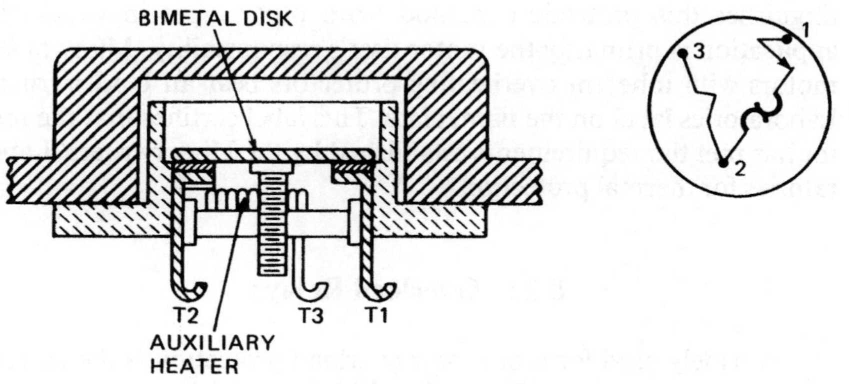


**Figure 8.16.** Inherent-overheating protector.

passes through a heater element which is located inside of the thermal protector.

This type of protector is usually mounted in the motor end shield so that it senses the motor's internal temperature plus the heat generated by its own heater. Veinott[3,4] has described the theory of this type of protector in the following manner.

The total temperature is the sum of three factors:

1. Ambient temperature
2. Rise of the end shield above ambient
3. Rise of the windings above ambient

Since the protector is located inside the motor end shield, it is directly exposed to a temperature which is the sum of items 1 and 2. The heater contained within the protector carries motor current and, therefore, the watts generated by the heater are directly proportional to the stator $I^2r$ losses. Consequently, the temperature at the bimetal disk in the thermostat varies in direct proportion to the actual winding temperature even under conditions of varying load and varying ambient temperature. Since the inherent protector's heater characteristics parallel closely the heating of the winding, the protector can be selected so that it will disconnect the motor from the line at any time the temperature reaches a preselected unsafe value.

This type of thermal protection is provided by the motor manufacturer as an integral part of the motor. The application of the protector is, therefore, performed by the motor manufacturer, which distinguishes this protection method from other types in which the application is primarily the motor user's responsibility. Many of the motors with inherent overheating protectors bear an Underwriters' Laboratories label on the nameplate. This label certifies that the motor has met the requirements established by the Underwriters' Laboratories for thermal protection.

## 8.11. Overload Relays

A widely used form of motor overload protection is the current operated overload relay. This type of relay is sensitive to both motor current and the time duration of the current.

There are two types of overload relays in common use. The magnetic type, shown in Fig. 8.17, consists of a coil wound around a sealed tube. Inside of the tube is a movable magnetic core. The motor current passes through the coil, and the coil's magnetic field attempts to center the magnetic core inside the coil. However, the movement of the iron core is resisted by both a spring and a damping fluid. As the magnetic core moves toward the center of the coil, the flux linking the coil increases until it reaches a threshold value and the contact opens. The purpose of the movable core is to provide a time delay in the flux buildup so that the relay exhibits a time–current characteristic, referred to as an "inverse-time" curve. The magnetic field buildup is a combination of the current in the coil and the increasing permeabil-

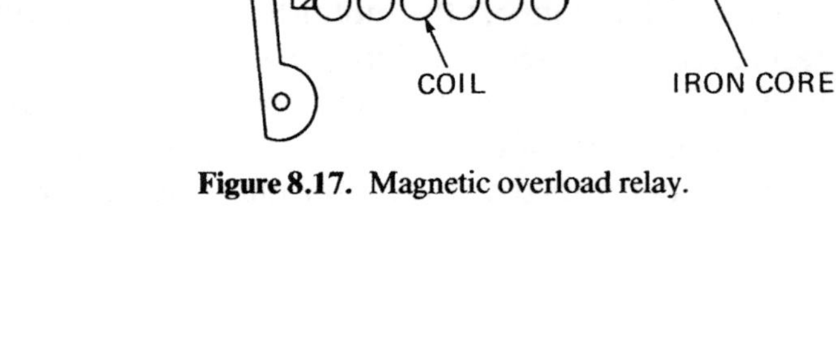


**Figure 8.17.** Magnetic overload relay.

ity as the moving core centers itself inside of the coil. At very high currents, approximately 10 times rated current, the magnetic field is sufficient to trip the relays instantaneously without waiting for the movement of the magnetic core.

Thermal relays are also used to provide inverse-time overload protection. The motor current passes through a heater element in a thermal relay and the heat thereby generated causes a heat-sensitive element to open at some predetermined temperature. The thermal storage capability of the relay elements provides the desired inverse-time characteristic. Since this type of relay stores heat, it will operate more rapidly on repetitive overloads than when the relay is initially at ambient temperature. The motor windings, of course, have the same characteristic and also experience temperature escalations on repetitive overloads, so that thermal relays duplicate the motor's heating curve within limits.

Figure 8.18 illustrates the application of an overload relay. The figure displays the inverse-time characteristic of an overload relay and illustrates that tripping time is relatively fast for large values of current and increases as the current approaches the motor's full-load value. For continuous operation, the overload relay is set to trip at some percentage value greater than the motor's full-load current. This percentage usually ranges between 115 and 125% depending on application conditions.[5] Also shown on the same graph is the motor's overheating curve. This curve is the current–time curve that, when exceeded, will result in overtemperature conditions in the motor. The overload relay should trip at some value less than the overheating curve as shown in the figure.

Figure 8.18 also displays the motor current versus time curve drawn for an acceleration from zero speed up to operating speed. This curve is, of course, dependent on the motor application conditions—load, $Wk^2$, and voltage—applied during the acceleration, so it is not a generic curve for a given motor and protector. However, the acceleration current curve can be drawn once the application conditions are known.

Figure 8.18 displays a successfully applied motor overload relay. The current versus time curve for the motor during acceleration lies beneath the overload trip curve, so the motor will accelerate without tripping the overload relay. Also, the overload relay tripping curve

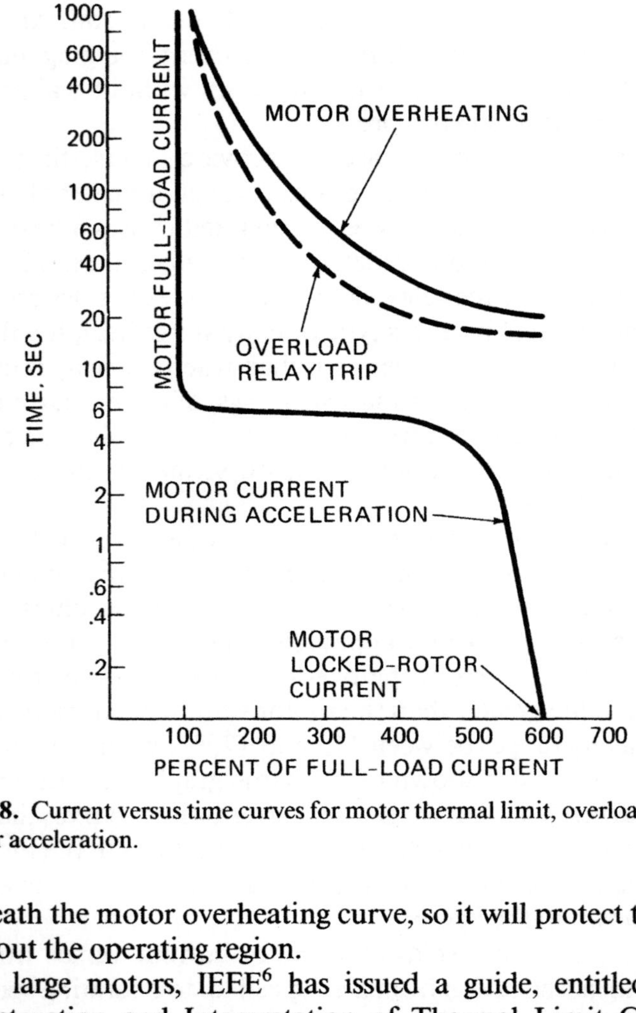


**Figure 8.18.** Current versus time curves for motor thermal limit, overload relay trip, and motor acceleration.

lies beneath the motor overheating curve, so it will protect the motor throughout the operating region.

For large motors, IEEE[6] has issued a guide, entitled "Guide for Construction and Interpretation of Thermal Limit Curves for Squirrel-Cage Motors Over 500 Horsepower," for constructing and using curves of the type displayed in Fig. 8.18. The protection of a large motor justifies the time required to obtain the motor's thermal limit curve and match it against the overload relay and acceleration–current curve. This procedure is frequently followed during the design phase of a project where large motors are involved. For smaller motors, overload protectors are usually applied by selecting the overload relay from the manufacturer's catalogs, based on the motor's rated current.

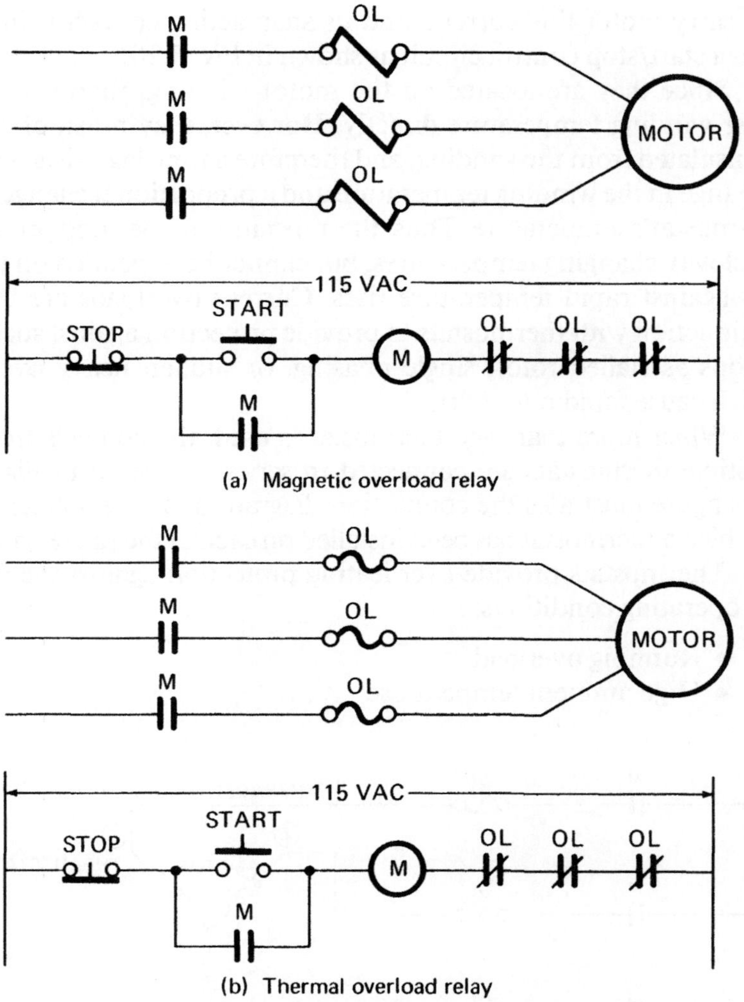


**Figure 8.19.** Circuit diagrams for magnetic and thermal overload relays.

Figure 8.19 shows circuit diagrams illustrating the use of magnetic and thermal overload relays.

## 8.12. Thermostats

Thermostats are hermetically sealed, snap-acting devices which are installed on the motor winding end turns. The thermostat does not carry motor line current, and its snap-acting contact is in series with a start/stop control circuit as shown in Fig. 8.20.

Since they are located on the motor winding, thermostats can sense winding temperature directly. However, they must, of course, be insulated from the winding, and therefore a time lag exists between a change in the winding temperature and a proportional change in the thermostat's temperature. Thus, thermostats provide good protection for slowly changing temperatures, but cannot be depended on to protect against rapid temperature rises. Current overloads are used in conjunction with thermostats to provide protection against such conditions as stalled rotor, single phasing, or sudden heavy overloads which cause rapid rates of rise.

When more than one thermostat is used, the contacts from the multiple thermostats are connected in series, as shown in Fig. 8.20. This figure illustrates the connection diagram of a three-phase motor in which a thermostat has been installed on each of the phase windings.

Thermostats provide overheating protection against the following operating conditions:

- Running overload
- High ambient temperature

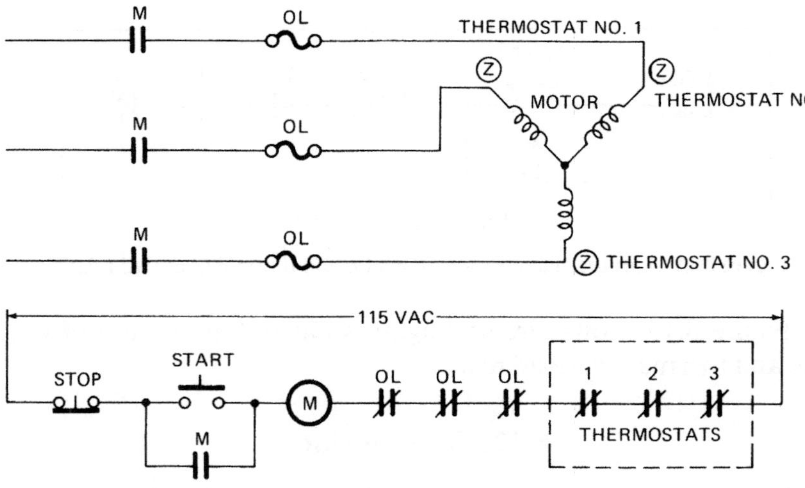


**Figure 8.20.** Wiring diagram for three-phase motor with current overloads and thermostats.

- Blocked ventilation
- Line voltage variation
- Unbalanced voltages
- Other conditions causing a gradual voltage rise

## 8.13. Embedded Temperature Detectors

Embedded detectors are defined by NEMA[7] as resistance temperature detectors or thermocouples, built into the motor at locations which are inaccessible after the machine has been built. The resistance temperature detector (RTD) is a commonly used embedded detector. RTDs are made in the form of a strip which can be inserted into the stator slot, along with the coils of a form wound coil. The RTD must, of course, be separated from the coil conductor by insulation so that the RTD is not at motor coil potential. However, since it is embedded in the slot with the coils, it can accurately sense the winding temperature.

The resistance of the RTD varies linearly with temperature, so that the motor winding temperature can be determined by resistance readings taken on the RTD. This type of temperature detection serves two purposes. First, it is used as a measurement method for determining that a motor meets its specified temperature rise on test. Since the RTD is located in the machine so as to measure the temperature at the hottest location, NEMA[7] allows a higher temperature rise when the measurement is by RTD for all motors except those over 7000 volts. The temperature differential between RTD temperature rises and rises obtained by measuring the motor winding resistance varies between 5 and 15 °C, depending on machine size and voltage rating.

In service, RTDs are used to provide an indication of motor temperature displayed so that the operator can observe the winding temperature during operation. Sometimes the RTDs are connected into a control so that an alarm is sounded or the motor is shut down in the event of overtemperature.

RTDs are installed at least one per phase, and it is not uncommon to have a motor equipped with several spares in case one fails. Since the practice of using embedded detectors is usually limited to machines with form wound coils, embedded detectors are usually applied to large ac motors designed for operation at 2300 volts and higher.

## 8.14. Solid-State Protection

Solid-state motor protection systems are now available, and these systems combine several features of motor protection into a single module. The inputs of motor current and sometimes direct temperature sensors are supplied to the solid-state control. The control circuitry then provides logic to interpret the operating conditions in a more sophisticated manner than can be obtained by a simple relay or a snap-acting thermal switch. For example, solid-state modules provide an adjustable time delay before tripping on overcurrent. This feature can be used to allow the motor to accelerate a high-inertia load or operate through a temporary overload condition. It may be difficult to anticipate these conditions and match relay characteristics to allow for them during the design phase of a project. By using a solid-state control, the needed adjustments can be made during start-up. Phase reversal, open phase, and unbalanced voltage protection are other examples of the types of motor protection that can be provided by solid-state controls.

The types of devices available for motor protection are myriad. Some of the more commonly used protective schemes have been discussed here, but the motor user has available a wide variety of devices from which to choose. Final selection must be made with a view toward the cost of downtime and motor replacement or repair.

## 8.15. Surge Protection

Motors are sometimes damaged by lightning and system-induced surges. Whenever motors are installed in areas subjected to these types of voltage surges, installation of surge protection equipment should be considered. Excess voltages can arise from the following three causes:

- Lightning

- Surges related to capacitor switching
- Surges related to insulation failure of other devices in the system

The devastating effects of lightning are known to all of us. Motors connected to overhead distribution lines are always vulnerable to lightning strikes. This is true even though there are intervening transformers. Damage can result from either a direct lightning strike or voltage surges induced from nearby lightning.

Power factor capacitors are sometimes installed by power companies at strategic points along the network. Switching of these capacitors on and off the line can cause voltage surges as the switching devices restrike in the course of interrupting the current.

Breakdown of insulation at some point in the system other than the motor itself can also produce switching surges on the transmission lines. Koerber[8] states that these types of failures can produce voltage peaks nearly three times the normal line-to-ground crest voltage value.

All of these occurrences cause steep wave fronts of magnitudes several times rated voltage to appear at the motor terminals. This has two consequences, both of which are undesirable. First, the voltage magnitude alone impressed at the motor terminals may exceed the design voltage. Second, the steep wave front concentrates this voltage across the first few turns of the motor's coils, which are connected to the line leads. According to Shankle *et al.*,[9] if the wave front is less than the electrical length of the windings as a whole, the full voltage will be dropped across the portion of the winding which has an electrical length that matches the wave front, usually a few turns. In this situation, failure occurs at the motor coils which are connected directly to the line.

The steep wave fronts which occur during a voltage surge cause the transmission line connected to the motor terminals and, in fact, the motor itself, to appear not as a simple circuit composed of resistances and inductances, but as a distributed network containing capacitance as well. A typical representation for a transmission line over which a voltage surge is traveling is displayed in Fig. 8.21. The values of $L$ and $C$ in Fig. 8.21 are usually expressed as henries per unit length, such as henries per foot, and farads per unit length. The parameter which characterizes such a transmission line is its surge impedance:

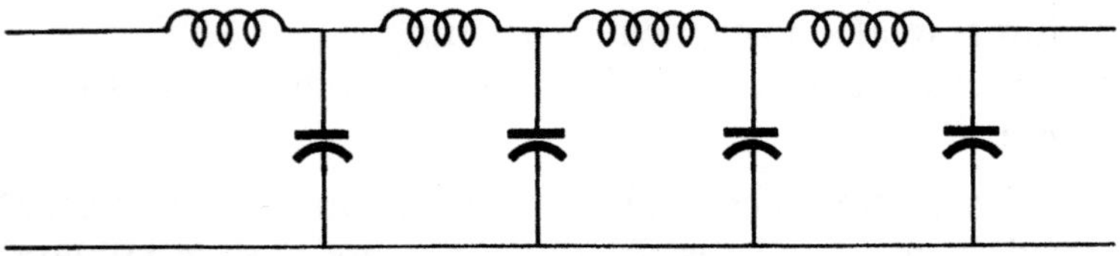

**Figure 8.21.** Equivalent circuit representing the distributed constants of a transmission line.

$$Z = \sqrt{\frac{L}{C}} \tag{8.1}$$

where $Z$ = surge impedance in ohms, $L$ = inductance per unit length, and $C$ = capacitance per unit length.

Another characteristic of a transmission line is that the incident voltage or current wave travels down the line at a velocity of

$$v = \frac{1}{\sqrt{LC}} \text{ unit lengths per second} \tag{8.2}$$

Motors connected at the end of the transmission line can also be characterized by a surge impedance value. A typical range of surge impedances for motors larger than 100 hp is 500 to 5000 Ω.[10]

Normally the surge impedance of a motor is so much larger than the surge impedance of the transmission line leading to the motor that the motor appears as an open circuit to the voltage surge traveling down the line. The surge is, therefore, reflected back down the line in the opposite direction with the same polarity causing an approximate doubling of the voltage magnitude at the motor terminals.

Walsh[11] suggests some equivalent circuits using lumped circuit parameters to approximate a distributed parameter transmission line which aid in understanding this phenomenon. These equivalent circuits are displayed in Fig. 8.22. For these circuits the equations relating the junction voltage to the incident voltage are given in Eqs. (8.3), (8.4), and (8.5). The incident voltage, $2E$, is the sum of the applied voltage, $E$, plus its reflection.

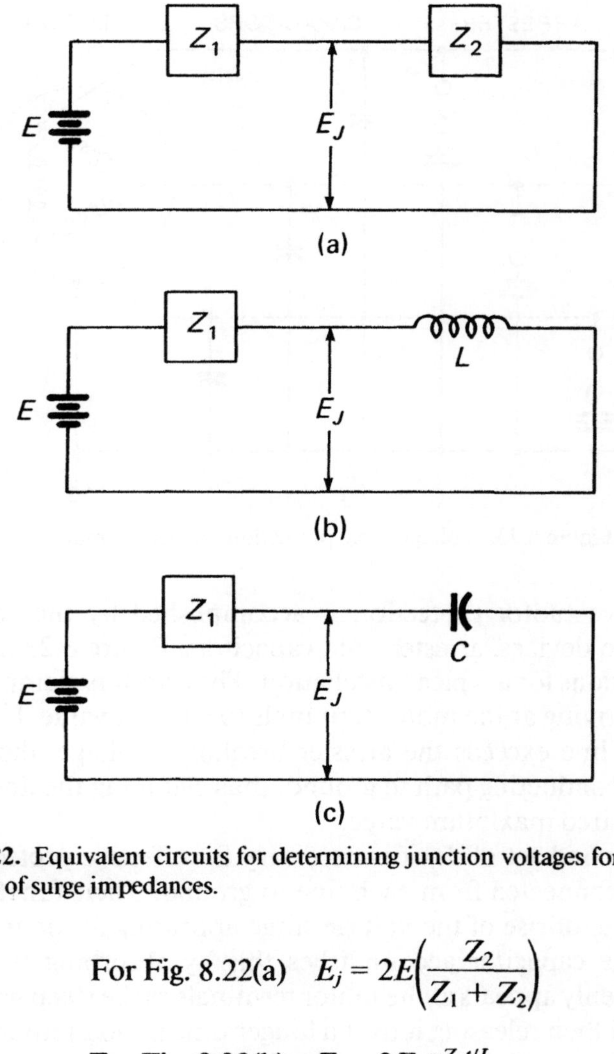


**Figure 8.22.** Equivalent circuits for determining junction voltages for different terminations of surge impedances.

$$\text{For Fig. 8.22(a)} \quad E_j = 2E\left(\frac{Z_2}{Z_1 + Z_2}\right) \tag{8.3}$$

$$\text{For Fig. 8.22(b)} \quad E_j = 2Ee^{-Z_1 t/L} \tag{8.4}$$

$$\text{For Fig. 8.22(c)} \quad E_j = 2E(1 - e^{-t/Z_1 C}) \tag{8.5}$$

Figure 8.22(a) is representative of a transmission line and a motor with no protective devices installed. If the motor's surge impedance $Z_2 \gg Z_1$, then the voltage at the motor terminals approaches twice the applied voltage, or $2E$.

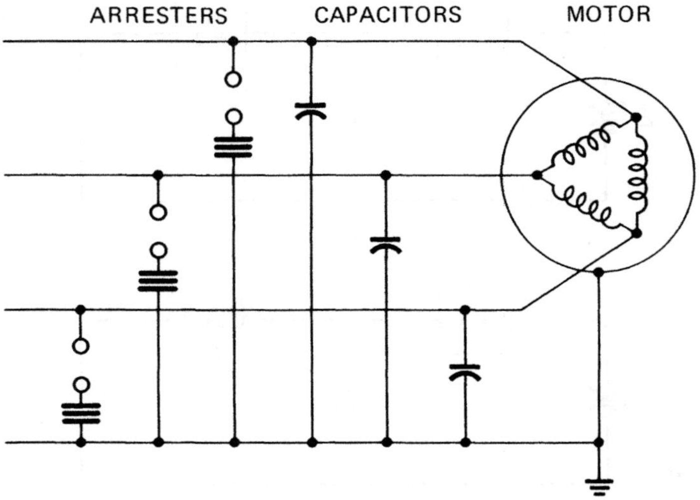


**Figure 8.23.** Voltage surge protection for electric motor.

Effective motor protection is accomplished by the combined effect of two devices, arresters and capacitors. Figure 8.23 illustrates the connections for a typical installation. The arresters clamp the voltage level arriving at the motor terminals to a preset value. If the voltage on the line exceeds the arrester breakover voltage, the arrester provides a conducting path to ground, thus clamping the line voltage to some desired maximum value.

The second set of devices necessary for effective protection are capacitors connected from each line to ground. Their function is to limit the rate of rise of the voltage surge appearing at the motor terminals. The capacitor accomplishes this by absorbing the energy which suddenly appears at the motor terminals as the steep wave front arrives, and then releasing it over a longer time period into the motor winding. The beneficial effect derived from capacitors is that the voltage surge is distributed more evenly across the entire winding rather than being absorbed by the first few turns as would be the case without the capacitors present.

Engineers working with power equipment are not usually concerned about lead lengths needed to interconnect various power devices, as they contribute negligibly to the impedance and losses of the complete system. However, in engineering a surge protection system, one must remember that voltage surges are a high frequency phenomenon, not 60 Hz power, and lead lengths become important. This can be illustrated by an example of a typical installation showing the effect of varying lead lengths between the motor terminals and the surge capacitors.

Figure 8.24 shows a 4000-V motor connected to a transmission line with $Z = 25\ \Omega$ and a surge voltage of 16 kV, wave front of 100 nsec, impressed on the system. Lead lengths connecting the motor terminals to the capacitors have the effect of adding inductance at the rate of 0.25 $\mu$H per foot in series with the capacitor. Any internal inductance of the capacitor itself is assumed to be included with the lead length inductance for this example.

The critical period from a voltage crest standpoint is the first 100 nsec during which the voltage is building up at the motor terminals. Assuming C = 0.5 $\mu$F, a typical value of surge capacitance, then only a small fraction of the incident voltage can build up across the capacitor in 100 nsec. Using Eq. (8.5), the voltage across C would reach only 255 V without any inductance present in the circuit connecting the motor terminals to the capacitor. However, with lead inductors present, the circuit appears as in Fig. 8.22(b) for the first 100 nsec. Equation (8.4) can be used to calculate the voltage appearing at the motor terminals. Since Eqs. (8.3) through (8.5) all assume a step voltage, zero front time, the 100 nsec rise time must be approximated by a series of smaller step voltages and the results added to obtain the

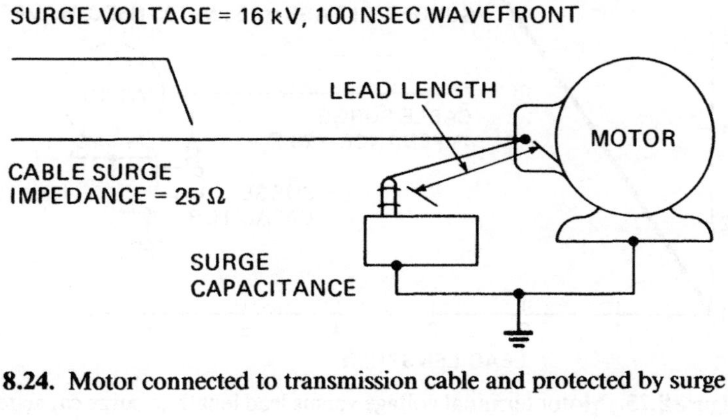


**Figure 8.24.** Motor connected to transmission cable and protected by surge capacitors.

final value of $E_j$. This technique suggested by Walsh was used to calculate the curve in Fig. 8.25 which shows the pronounced effect that long lead lengths have on counteracting the beneficial effects of surge capacitors. For example, with 1-foot leads, the motor terminal voltage is limited to 3.2 kV, but if 5-foot leads are used, the terminal voltage will reach 13.8 kV.

A similar phenomenon occurs with lightning arresters, and lead lengths to arresters should also be kept to a minimum.

The application decision on whether or not to apply surge protection must be made on an individual basis. The costs of repairing or replacing the motor if damaged, and the additional costs of being out of service, must be weighed against the cost of the surge suppression

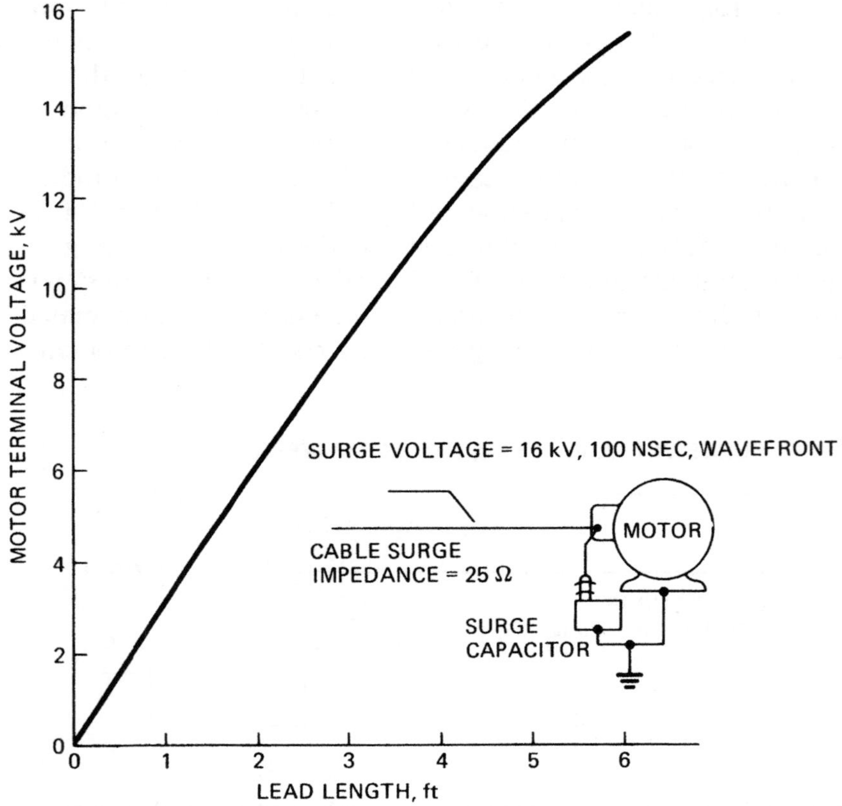


**Figure 8.25.** Motor terminal voltage versus lead length to surge capacitors.

equipment and the probability of voltage surges occurring. Large horsepower ratings for use on 2300-V and higher systems are the ones most commonly selected for application of surge protection equipment.

## References

1. NEMA Standards Publication No. ICS2, "Industrial Control Devices, Controllers and Assemblies." Washington, D.C.: National Electrical Manufacturers Association, 1988.
2. Courtin, J.J. Ten part-winding arrangements in sample 4-pole induction Motor, *IEEE Transactions, Part III, Power Apparatus and Systems,* Vol. 74, 1955.
3. Veinott, C.G. *Theory and Design of Small Induction Motors.* New York: McGraw–Hill Book Co., 1959.
4. Veinott, C.G., and Schaefer, L.C. Fundamental theory of inherent overheating protection under running overload conditions, *AIEE Transactions, Power Apparatus and Systems,* 1949.
5. National Electrical Code, National Fire Protection Association, Boston, 1981.
6. "Guide for Construction and Interpretation of Thermal Limit Curves for Squirrel-Cage Motors Over 500 Horsepower," *P620,* IEEE, New York, 1981.
7. NEMA Standards Publication No. MG1, "Motors and Generators," Washington, D.C.: National Electrical Manufacturers Association, 1993.
8. Koerber, A.R. Surge protection for rotating electrical equipment, *Plant Engineering,* Sept. 7, 1972.
9. Shankle, D.F., Moses, G.L., and Mole, C.J. Steep-front surges endanger windings of powerhouse motors, *Electric Light and Power,* November, 1966.
10. Andrä, Walter and Sperling, Paul-Gerhard. "Winding Insulation Stressing During Switching of Electrical Machines," *Siemens Review,* XLIII, No. 8, 1976.
11. Walsh, G.W. A review of lightning protection and grounding practices, *IEEE Transactions on Industry Applications,* Vol. IA-9, No. 2, March/April, 1973.

# Index

9 780306 446986

GPSR Compliance

*The European Union's (EU) General Product Safety Regulation (GPSR) is a set of rules that requires consumer products to be safe and our obligations to ensure this.*

*If you have any concerns about our products, you can contact us on ProductSafety@springernature.com*

In case Publisher is established outside the EU, the EU authorized representative is:

Springer Nature Customer Service Center GmbH
Europaplatz 3
69115 Heidelberg, Germany

**Batch number: 10755292**

Printed by Printforce, the Netherlands